The Pioneering Role of
CLARENCE LUTHER HERRICK
in American Neuroscience

BOOKS BY WILLIAM FREDERICK WINDLE

An Outline and Laboratory Guide to Neurology
Physiology of the Fetus: Origin and Extent of Function in Prenatal Life
Textbook of Histology (with J. F. Nonidez), First Edition, 1949
          Second Edition, 1953
Asphyxia Neonatorum
Bacterial Pyrogens: An Annotated Bibliography (with L. G. Ginger and
     Ione E. Johnson)
Textbook of Histology
          Third Edition, 1960
          Fourth Edition, 1969
          Fifth Edition, 1976*
Physiology of the Fetus: Relation to Brain Damage in the Perinatal Period†
The Spinal Cord and Its Reaction to Traumatic Injury
The Pioneering Role of Clarence Luther Herrick in American Neuroscience

AS EDITOR

Regeneration in the Central Nervous System
New Research Techniques of Neuroanatomy
Biology of Neuroglia
Neurological and Psychological Deficits of Asphyxia Neonatorum
The Process of Aging in the Nervous System (with J. E. Birren and H. A.
     Imus)
Neural Mechanisms of the Auditory and Vestibular Systems (with C. L.
     Rasmussen)

*Spanish edition
†Spanish and Italian editions

C. L. Hemck

The Pioneering Role of
CLARENCE LUTHER HERRICK
in American Neuroscience

William Frederick Windle
Denison University, Granville, Ohio

An Exposition-University Book

Exposition Press   *Hicksville, New York*

Parts of this book appeared in *Experimental Neurology,* vol. 49, 1975, and are reproduced with permission of Academic Press, Inc.

First Edition

Library of Congress Catalog Card Number: 78-75284

ISBN 0-682-49340-6

Printed in the United States of America

CONTENTS

PREFACE

This book describes some episodes in the life of Clarence Luther Herrick that relate to his pioneering role in development of neurobiology in America. Herrick was a versatile man whose writings ranged over the fields of geology, zoology, and moral philosophy as well as psychology. Others have evaluated the significance of his contributions in those fields. William Tight told about his work on the geological formations in Ohio and New Mexico. Henry Bawden summarized his metaphysical contributions. Charles Judson Herrick twenty-four years ago published a biography of his oldest brother in which he emphasized the latter's role as a psychobiologist. Today, in the light of our concept of neuroscience, it is clear that Clarence Herrick qualifies as the pioneer in that field, for he was the first to propose an integration of all aspects of investigation of the brain in order to understand the mind. He attempted to establish a laboratory of neuroscience, but in that he failed, for the time was not yet right.

Herrick came onto this stage at the time when scientists were developing new methods to study structure and function of tissues and organs of the body. Emergence of experimental biology from natural history was underway. Mysterious parts of the brain to which early anatomists had given fanciful names (e.g., olive, pons, hippocampus, or cauda equina) were yielding their secrets to a new corps of experimentalists, especially in Europe. He began to qualify in that category a century ago with his youthful essay on embryology of the chick. A few years later in Leipzig he came to realize that psychology, purporting to deal with the mind, should have to incorporate anatomy and physiology of the brain if it were to emerge from its position in the realm of a philosophy to that of a true science. One of his greatest contributions was

7

made there; from it arose an abiding urge to fathom the human mind or soul.

To understand the man it is necessary to envisage the boy. This was perhaps the most difficult task because no one who knew him in those early formative years left any records of their memories. Consequently, I have resorted to a certain amount of speculation based on accounts of life on the frontier in those times.

No doubt Clarence Luther Herrick's character had been shaped before he was ready to enter high school. The deeply religious background provided by his father, a Free Baptist minister, had a profound influence upon it. The loving devotion of an intelligent mother set a cultural pattern. His childhood in isolation on the frontier gave him freedom to acquire an intimate relationship with nature. Self-reliance and sense of responsibility developed in response to rugged life on the farm. His role as pioneer neuroscientist emerged from such a background.

The record of Herrick's life during the last two decades of the nineteenth century is well documented. Not only are there his published writings, but many letters and notes have been preserved. Without the latter it would have been impossible to form a clear picture of the man and his concept of neuroscience. Most of these are in the Special Collections Division of the Kenneth Spencer Research Library of the University of Kansas at Lawrence, while others are in the archives of the William Howard Doane Library of Denison University at Granville, Ohio. Documents in the archives of the University of Chicago were thoroughly abstracted by Professor Lincoln Blake for a 1966 doctoral dissertation. Other records of Herrick's life have been kept by the University of New Mexico, where he served as president during some of his years of attempted convalescence from tuberculosis.

Charles Judson Herrick's biography of his older brother was the source of much information. That biographer's deep affection and reverence for his brother gave his account a personal quality that could not be achieved by another writer. It was partly autobiographical, and Charles Judson Herrick tended to overlook a number of episodes that did not cast the life of his brother in the most favorable light. He was ten years younger, and remembered his brother best from the years in which a close professional association developed, particularly from the time when Clarence became dependent on him because of the tragic illness that incapacitated him. Intimacy was mainly lacking at earlier times when Clarence was his taskmaster and teacher.

I have covered the University of Chicago episode in more detail than did his brother. Herrick's intense bitterness over it affected his efforts in neuroscience and was a factor in the breakdown of his health. The tragic ruination of a sensitive life was evident from the many letters that survive. His retreat into the realm of religious philosophy resulted from the discovery that the word of even the most exalted men could not be trusted and that of God was doubted. So ended his life.

I am grateful to a number of people who helped me find the material upon which this account is based. Access to the Herrick files at the University of Kansas was provided by Paul G. Roofe and Alexandra Mason; Margaret Hanson, at Denison, made many of the contacts with other libraries; John N. Durrie, secretary of the University of New Mexico, provided an extensive list of material available at Albuquerque; Maxine B. Clapp, archivist of the University of Minnesota library, and Alice M. Vestal, of the Special Collections Department of the University of Cincinnati Library, located a number of important items; and Lincoln C. Blake, of Earlham College, lent a copy of his thesis. The manuscript was read critically by Milton Emont, Horace Magoun, Louise Marshall, Paul Roofe, and Ella Windle, to all of whom I extend thanks.

Publication
of this biography was sponsored
by the Denison University Research Foundation,
aided by a gift from Dexter J. Tight.

The Pioneering Role of
CLARENCE LUTHER HERRICK
in American Neuroscience

I

BACKGROUND OF
AMERICAN NEUROSCIENCE

Neuroscience is a relatively new term applied to interrelated studies in the nervous system, i.e., the brain, nerves, neuro-muscular apparatus, and sense organs. It embraces not only neuro-anatomy, neurophysiology, and neurochemistry, but also some aspects of endocrinology and much of the psychological, com-municative, and behavioral sciences. Although the term generally refers to research and teaching activities in the basic sciences, human clinical studies are by no means excluded because interest in the nervous system developed in America largely from the branch of medicine formerly called nervous and mental disease. The recognition of a need to accomplish integration of disciplines is clearly indicated by the organization recently of a Society for Neuroscience, representing the culmination of a movement that has been underway for a considerable period of time.

Research on structure and function of the nervous system was advancing rapidly in Europe during the second half of the nineteenth century. The nucleus of a neuroscience was in evidence there before it could be said to exist elsewhere. Brown-Séquard in 1849, for example, described experiments on pigeons in which he severed the spinal cord, studying the effect on their movements and sensation and the histological changes occurring at the site of the cut.[1] Fritsch and Hitzig did their famous experiments in 1870, stimulating the exposed cerebral cortex of dogs and inducing contractions of muscles of the legs.[2] In the 1880s, the great Spaniard Santiago Ramón y Cajal was embarking on his career in neuroscience that led to a Nobel Award (shared with Camilio Golgi) in 1906. The last two decades of the nineteenth century saw great progress in European research in the nervous system.

When and where did the movement toward neuroscience commence in America? And can one say who led it? The answers to these questions appear to be: during the 1880s; in Ohio; by Clarence Luther Herrick, naturalist, philosopher, teacher, artist, and scholar of rare genius. He was the pioneer American neuroscientist, not the first to study brains, but the first to propose exploration of interrelations between neural structure and function, comparative and integrative, in a search for understanding the mind of man. Others before him had done research in the nervous system. He was the one, however, who saw the necessity of bringing together physiological and histological findings and the need to compare the brains of various species. And he was among the first to try to lift psychology out of the esoterism of the metaphysician, and bridge the dichotomy of body and mind.

His name is relatively unknown to American neurologists today. Strange as it may seem, it was not included by compilers of the two principal collections of neurological biographies.[3,4] It is time to recognize his pioneering in neuroscience and put his name in its proper place among those of his peers.

Herrick became interested in the nervous system while doing graduate work at Leipzig in 1881 and 1882. At that time there were few other Americans of his age who were publishing articles in that field, and most of them had other primary interests.[5] A number of American scientists about ten years younger than Herrick were soon to become known for their neurological research.[6] Nearly all research on the nervous system was done by men who had been trained in European laboratories or who were pupils of investigators who had studied there.[7]

All but a small number of scientific articles on brain or nerves flowed from institutions of Europe in the 1880s. The only American neurological journals when Herrick entered the field were the *Chicago Journal of Nervous and Mental Disease,* first published in 1874, and the *American Journal of Neurology,* in 1881. They were clinical. American periodicals of general science occasionally carried articles on the nervous system. Among them were the *Scientific American, American Naturalist, Popular Science Monthly,* the *Microscope,* and *Science.* The *Journal of Morphology* began publication in 1888. Herrick used some of these. Investigators in that period usually published in general medical journals and in the proceedings of academies, as there was no periodical specializing in neurobiology in any country.

There was little communication among the few American neurobiologists of the 1880s. They did not go to distant meetings of scientific societies, for indeed there were few societies that held meetings except the American Association for Advancement of Science, formed in 1883, and various medical societies. The American Neurological Association had been organized in 1875 by clinical neurologists. The Association of American Anatomists began to meet in 1888, one year after the American Physiological Society. The First International Congress of Physiological Sciences was held in Basel, Switzerland, in 1889, but no American neuroscientist attended it.

Early interest in neurobiology was physiological rather than anatomical. Two American centers were located at Harvard Medical School in the laboratory of Henry P. Bowditch, and at the new Johns Hopkins University in Baltimore where H. Newell Martin organized a laboratory for experiments in physiology including neurophysiology. Reports on research in the nervous system came from both of these places, but neurophysiology was not the sole interest of any of the scientists trained in those institutions prior to the 1880s.

Neuroscience made a small beginning during that period. A number of men in this country contributed to its development, each in his own limited field of interest. Clarence Luther Herrick's vision of an integration of neurological disciplines was unique; although his plan to establish a department of neuroscience failed to become effective in his lifetime, it marks him as the founder of neuroscience in America.

NOTES

1. C. E. Brown-Séquard, Expériences sur les plaies de la moelle épinière, *Gaz. Med. Paris Ser. III,* 4: 232-33, 1849.

2. G. T. Fritsch und E. Hitzig, Uber die elektische Erregbarkeit des Grosshirns, *Arch. Anat. Physiol. wiss. Med.* Jg. 1870: 300-332.

3. L. C. McHenry, Jr., *Garrison's History of Neurology* (Springfield, Illinois: Charles C Thomas, 1970).

4. W. Haymaker and F. Schiller, *The Founders of Neurology,* 2nd ed. (Springfield, Illinois: Charles C Thomas, 1970).

5. Contemporary scientists with neurological interests were: E. P. Allis, N. E. Brill, W. Browning, H. H. Donaldson, C. F. Hodge, W. H. Howell, F. W. Langdon, W. A. Locy, H. N. Martin, A. D. Morrill, H. F. Osborn, W. T. Sedgewick, H. Sewall, E. L. Spitzka, T. B. Stowell and B. G. Wilder. See "The Neurosciences," by Frank, Marshall & Magoun, 552-613, in *Adv. Amer. Med.,* New York, Josiah Macy, Jr. Found., 1976.

6. Younger men entering the field were: J. Hardesty, R. G. Harrison, C. J. Herrick, G. C. Huber, C. E. Ingbert, J. B. Johnston, B. F. Kingsbury, H. McE. Knower, A. Meyer, G. A. Parker, S. Paton, D. A. Shirres, C. A. Strong, and O. S. Strong.

7. Interest in the nervous system was stimulated by the writing of European scientists, notably Bell, Bernard, DuBois-Reymond, Ferrier, Golgi, Helmholtz, His, Hitzig, Kolliker, Magendie, Muller, Remake, Schwann, and Waller. They and others had established neurology on a firm scientific basis. The transient presence in the United States of Brown-Séquard prior to 1878 greatly influenced development of medical neurology.

II

BOYHOOD ON THE FRONTIER

Clarence Luther Herrick was born on June 22, 1858,[1] near the village of Minneapolis, Minnesota, a region that was still part of the American frontier. The land had been ceded by the Sioux Indians to the federal government only six years earlier and its settlement was taking place with remarkable rapidity, influenced by the proximity of St. Anthony Falls on the upper reaches of the Mississippi River, which provided waterpower for mills. The Herrick homestead was located about three miles from the falls—within earshot of them. The village population by that time had grown to several thousand; churches and schools had been built and a public library established in this thriving community. Minnesota became a state the year Clarence was born.

Clarence's grandparents had migrated with their four grown children from Stowe, Vermont, in 1854. Their first stop was Dubuque, Iowa, where they tried to establish a family business, manufacturing tombstones. Henry Nathan Herrick, Clarence's father, was then twenty-one years old. Four years later Henry's parents decided to move on to Minnesota to acquire land, but before that happened Henry left them and returned to the East. He aspired to become a physician and when he reached New York City he entered an eclectic institute of hydrotherapy, not realizing its fraudulent nature.[2] Although he finished the course and obtained a diploma, he disdained to use it and decided instead to go into the ministry.

He married Anna Strickler, a schoolteacher in the city of Washington, in 1857.[3] Anna's home was Clarence, New York, where the marriage took place. (She gave the village name to her first-born son.) After the marriage the couple visited Niagara Falls and then turned to the West. In Minnesota they took up residence in a cottage on five acres of his father's land, where Henry strove to take a living from the soil while preparing himself for the work of God. He was ordained as a Free Baptist missionary minister in 1862. Clarence was four years old when his father received his first pastorate.

Clarence Herrick's life on the Minnesota frontier strongly influenced his future. As an only child for several years, he experienced a closer relationship with his mother than did his younger brothers. This must have been an important factor shaping his character. The deeply religious background of the Herrick home also played an important part. He became a scientist but remained an earnest Christian.

The five-acre tract of land upon which Clarence's father had built their first cottage was located on the outskirts of the Minneapolis settlement, surrounded by virgin prairie, lakes, streams, and wooded ravines. Five fertile acres were just about adequate to provide a subsistence living—one that today would be considered well below the poverty level. It did nonetheless provide most of the food required by the family of three. On it they grew vegetables and raised chickens. There was ample forage for a horse and cow, prairie grass being inexhaustible. Seldom was there meat on the table, but milk, butter, and eggs were plentiful, and presumably game provided an occasional supplement.

Money to buy clothing and staples, such as flour and sugar, was scarce and may have been earned by laboring in the rapidly growing village. Even in the year after Clarence's father became a minister he received no more than fifty dollars from one congregation and sixteen from another. There were few wealthy people in the area at that time, and austerity was the accepted lot of uncomplaining pioneers.

Although the Herrick home had none of the luxuries we take for granted as essentials of a well-ordered life today, it was not without some cultural amenities. Both of Clarence's parents were literate people. Without academic backgrounds, they nevertheless possessed a homely culture well above that of many other pioneers. Books were available from the Athenaeum, a public library that had been started in Minneapolis in 1857. Clarence's mother read to her son, and it was not long before he learned to read. A small portable melodeon, or reed organ, later obtained for use in church programs, was kept in the house, and with his mother's help Clarence learned to pick out tunes on it, eventually becoming quite an accomplished musician.

It is difficult to imagine the Minnesota environment in the 1850s. It was rugged. Winter brought long periods of quiet darkness punctuated only by calls of coyotes and other beasts. Hard work was required to keep the domestic animals and themselves alive and healthy, and each member of a family had chores to perform. Water had to be carried and the woodpile provided fuel to heat it. There was only one lamp in the Herrick household and all nighttime activities centered around it.

With spring came relief from the bitter cold, and release to the out-of-doors. Ice on the river broke and the drone of the falls resumed. All around were the sounds of awakening life: songs of birds and humming of insects soon filled the air. The country was rich in living, growing things that attracted a lonely youngster on long summer days.

As a child, Clarence was alone much of the time, for he had no brothers or sisters during his early years.[4] There were no neighbor children to play with. Organized activities for preschool boys and girls did not exist, and only on Sundays at the church in the village did he have opportunities to associate with others.

Isolated as the family was, there were few events to divert a child's attention from nature. Clarence had to amuse himself most of the time and, with few toys, there was little to do but play with objects and small creatures that he encountered near the house. No wonder that he formed a close relationship with wild life, one that led him into the study of biology as he matured.

Not every boy growing up under similar circumstances became so dedicated to science. It was obvious that there was more to it than his contact with nature. Clarence was an alert and active boy endowed with an unusually receptive mind. He possessed some remarkable innate abilities that began to manifest themselves quite early. Manual dexterity was one. He soon learned that he could use his hands—both of them—in response to things he saw, to draw their images on a slate or paper when available and, years later, for etching lithographs. He developed considerable artistic talent. Preference for drawing pictures instead of doing sums led to punishment in school on at least one occasion. His complete ambidexterity may have been associated with a lack of hemispheric dominance, a factor that could have been related to development of his genius.

Clarence's self-reliance and sense of responsibility were engendered during childhood. He was alone with his mother between the ages of six and eight. His father had volunteered for military service but was not called up until the spring of 1864. When the call came, he literally left the plow in the furrow and hastened to St. Louis to join his company. A commission was later awarded to him, and he spent most of two years as chaplain with the Union Army in the South. Clarence had to help his mother with farm chores while his father was away at war. I assume that she completed the spring plowing, and perhaps Clarence helped put seeds in the ground. They managed well, but it was months before his father's military pay began to arrive in Minneapolis, months of hardship when at times there was no bread on the table. Anna wrote to her husband: "I am out of wood and flour. I feel as if I could hardly keep house alone this summer."

They were too proud to complain, but they must have received help from other members of the church when the need became apparent. Henry Nathan Herrick certainly departed with complete faith in friends and neighbors who remained behind. Trust in one's fellowman was an attribute of the pioneer, second only to trust in God, which was especially strong toward the end of the Civil War.

Clarence Herrick at six walked to an ungraded district school in the village and began to receive a formal education consisting, no doubt, of learning the three Rs. But actually his education had begun at home where his mother, a former schoolteacher, was his tutor. She read to him by the light of the one coal oil lamp after evening tasks were done, when she could lavish upon him her undivided attention. She was of German extraction (had she for that reason named him Luther?), and I suppose she sang him German songs. Clarence soon began to pick up German words which eased his eventual task of acquiring fluency in that language, an accomplishment not attained by his brothers.

By the time Clarence was twelve years old he was shouldering major responsibility for the farm work. His brothers, William and Charles, were then four and two and required most of their mother's attention. His father left him in charge when he went to England in 1870 as guest of a group of businessmen of his congregation. Clarence was a tall lad with a rugged constitution and was fully capable of managing alone.

About that time his interest in nature passed beyond the stage of childish curiosity. He began in earnest to collect and preserve specimens and to draw them. His father's army rifle had been converted into a muzzle-loading shotgun with which he learned to bring down birds to study. He was joined in these activities by several other boys with whom he explored the wilderness surrounding Minneapolis.

The importance of his early years in Minnesota cannot be overstressed. By adolescence, Clarence Luther Herrick's character was shaped by his childhood experiences on the frontier, by freedom to acquire an intimate relationship with nature, by loving devotion of an intelligent mother, and by the trust in God and fellowmen engendered by a deeply religious father. Only the trust in the word of his fellowmen was to fail him and lead to a tragic end.

NOTES

1. He gave the date as June 21 in *Who's Who in America*.
2. A therapeutic cult called hydropathy was founded by Vincenz Priessnitz, a Silesian peasant. New York City had a "hydro" in 1853, operated by a "Dr." R. T. Thrall.

3. A previous marriage in 1854 had resulted in tragedy; his bride died of cholera on their journey to Dubuque. Anna Strickler three years later, answered his advertisement for a wife—on a dare, it was said.

4. A brother born in 1859 died in 1862.

III

YOUNG NATURALIST

Clarence Herrick entered the Minneapolis High School when he was sixteen years old, in the autumn of 1874, and began the standard course of study, including classical language, rhetoric, and mathematics. There were no classes in natural science, and biology generally was not taught in public schools at that time. Few colleges and universities even provided instruction beyond elementary botany and geography. Clarence had acquired a genuine interest in nature study and did not let the lack of course work prevent him from developing it further.

He and three other youths had spent part of the summer before entering high school camping on an island in one of the neighboring lakes. There they might have loafed or behaved like savages in the complete freedom of the wilderness, but chose instead to become amateur naturalists. They spent their time collecting and trying to identify the plants and animals of the island, especially the birds. That summer's experience marked the start of Clarence's serious interest in science. It led the boys to form their Young Naturalists' Society the following winter. His close friend, Thomas Roberts (who later became a physician and the director of the Minnesota Museum of Natural History), was one of the campers, and it was he who proposed the idea of a club.

The Young Naturalists' Society was launched by these and several other high school youths one evening in February, 1875, in the Roberts home. Years later (1939) Dr. Roberts described the organization in a letter to Clarence's brother Charles Judson Herrick:

Clarence did the best work and was indeed the only real scientist in the group. I think I was the promoter and furnished the place of meeting—my bedroom. This was on the second floor, reached

25

by a side entrance, through the dining room and up a back stairs. It was heated by a little drum stove and provided with storage arrangements for specimens, etc. Here we assembled and sat around in earnest conclave, for we were a serious bunch. Our youthful friends of different minds were wont to poke fun at us by sending us bogus specimens from time to time by way of joke. But we went gravely on our way. . . . Like most embryo students of nature we were regarded by the average run as a little queer— "odd fish." And I guess we were, as boys go.*

The topics discussed by the young naturalists ranged over the broad field of natural history. Although young Roberts's interest was mainly in birds, that of Herrick encompassed zoology, geology, and paleontology, as well as ornithology. Titles of a few of the reports are noteworthy in relation to his later attempt to organize studies in neuroscience: "Have Birds and Animals More than Instinct?"; "Localization of the Functions of the Brain"; and "Transmission of Excitations in Sensory Nerves (Paul Bert)."[1] Herrick's treatment of brain structure from the developmental viewpoint some years later was foreshadowed by a presentation entitled "On the Embryology of the Common Hen written from personal observations," which was recorded in the minutes of the Young Naturalists' Society but not published.

The meetings of the Society were held regularly until the end of the high school period. The minutes are filed in the archives of the Minnesota Museum of Natural History in Minneapolis, and a detailed discussion of them was published in 1942 by Charles Judson Herrick,[2] who commented on the significance of this organization:

The history of the Young Naturalists' Society is a record of an experiment performed for us unwittingly by a few sixteen-year-old boys who without encouragement or guidance beyond their own circle systematically organized their interests into an educational program which was internally motivated. The formal rules of procedures were conventional, but the design and actual operation of the educational machinery were their own.

---

*Errors in spelling and some of the punctuation in quotations throughout this book have been corrected.

Although there is no evidence that these young men received help or even guidance from high school teachers, it is not quite true that they lacked encouragement. Clarence's father certainly responded to his son's interest in biology and supported him to the limit of his ability to do so. After Clarence had entered high school the senior Herrick bought him a microscope similar to that shown in the photograph facing page 65.[3] This was a simple instrument, consisting essentially of two brass tubes, one sliding into the other, with objective lens at one end and eyepiece at the other. There was a mirror to direct light onto a specimen stage below the objective lens. The microscope had no condensor nor mechanism for fine adjustment. It cost eight dollars.

The microscope opened a whole new field of biology, that of creatures invisible to the naked eye, and Clarence lost no time beginning to study animalcules of pond water. Indeed, his first scientific article was an illustrated note, "A New Cyclops," read before the Young Naturalists' Society on November 25, 1875, and published in the *Fifth Annual Report of the Geological and Natural History Survey of Minnesota for 1876*. Clarence was then only eighteen years old. Another brief article, "Ornithological Notes," appeared in the same volume, and a third, "The Trenton Limestone at Minneapolis," came out in the *American Naturalist* the following year. The breadth of his catholic interest in natural history at this early period is revealed in the three titles.

Clarence spent only one year in the Minneapolis High School, after which he transferred to the preparatory department of the University of Minnesota. He donated a collection of more than one hundred bird skins to the high school as a parting gift—perhaps with an evangelical hope of encouraging other students to study nature. He continued to be active in the Young Naturalists' Society and intensified his field studies, especially those on microscopic animalcules.

According to his brother, Clarence's answer to a self-raised question: "Why should anyone devote such long hours of hard work to natural history as an avocation?" was: "We are conscious of a good result to ourselves from such study. We are also conscious of great pleasure in the study itself and can only account for it by supposing that there is an inherent desire in us that is met by this occupation." This is a surprisingly good statement of motivation of research from the pen of a high school freshman.[4]

His father accompanied him to the university in September for an interview with the president, and Clarence was immediately accepted as a "subfreshman" for a six-year course of study. Most of his friends remained in the local high school. The meetings of the Young Naturalists' Society continued for two more years and must have been enlivened by the reports of experiences that Clarence brought to them from his classes at the university.

The transfer to the university was a fortunate move. William Watts Folwell, the first president of that institution, is said to have been "a staunch and unintimidated champion of the scientific spirit."[5] He had studied in Berlin just before the Civil War and was developing a revolutionary plan of education, including in it the concept of a university in Minneapolis as a "federation of professional schools." He had outlined this in his inaugural address in 1869, and it became known as the "Minnesota Plan."

The University of Minnesota started with about three hundred boys and girls, half of them in the preparatory Collegiate Department, which was popularly known as the "Latin School," but a few were so ill prepared as to require tutoring in basic arithmetic and English grammar. By the time Clarence Herrick entered, enrollment in the university had declined and President Folwell was glad to have him.

Classes occupied relatively little time. They were not held on Monday and were confined to morning hours on other days. Each day began at eight o'clock and there was a compulsory chapel service about midmorning. Afternoons were free for study, work in laboratories, or reading in the college library.[6]

The cost of education at the university was negligible; tuition was free but there was an annual fee of five dollars for incidental expenses. Board and room in town could be had for two or three dollars per week, which Clarence saved by living at home. Thus, Clarence's annual cost of a college education amounted to only five dollars. His desire to transfer from the high school where teaching was conducted in the traditional method, to the university where there was considerable freedom to pursue his outside interests is understandable.

Clarence's interest in natural history soon came to the attention of Professor N. H. Winchell who, recognizing the neophyte's unusual aptitude and his skillfully executed drawings, obtained an appointment for him as his assistant for the following year. Winchell was professor of Geology and Mineralogy, State Geologist, and director of the Geological and Natural History Survey of Minnesota, which had been established in 1872 as an affiliate of the University of Minnesota.[7] The assistantship provided Clarence a small stipend, which was welcome, though no doubt he would have grasped the opportunity the position offered even had there been no financial reward.

One activity of the Society was to make collections and accurate lists of Minnesota flora and fauna. Another was to prepare essays on scientific topics, based largely on reading in the public library, and to present them at their fortnightly meetings. Some of these written papers contained original observations, or at least ones that the members believed to be original, and a few of the essays were later published.

Clarence Herrick entered the university with the expectation that six years would be required to complete the requirements for a bachelor of science degree, but the course schedules were re-organized and he had four years in the Collegiate Department and one year in the College of Science, Literature and the Arts. The faculty of the latter was composed of President Folwell and nine professors. There were only eight other students who completed the senior year and graduated with Clarence. The science course curriculum included chemistry, astronomy, physics, botany, zoology, geology, mineralogy, mathematics (through calculus), and surveying, all of which I assume he took.

The free afternoons were not always used for study in preparation for classes. Many of them and some weekends were devoted to field work with Professor Winchell. Then, too, there were home chores that could not be neglected but had to be done in the early morning hours, after which Clarence walked four miles to the

university. With all his extracurricular activities, formal classes
must have suffered, for he did not make the highest grades.
Clarence was only a "B" student with a general college grade
average of 82 percent.[8] However, if Professor Winchell gave
grades for the field studies, they surely must have been recorded
as "A."

There is no reason to believe that Clarence's college years
were filled exclusively with activities associated with natural
history. Some time was spent in work of his father's church.
Moreover, he taught school on one occasion, skipping one term in
his junior year to fill a vacancy in the local district school.

Favored with an outgoing personality, Clarence maintained an
interest in current events and developed his own opinions about
them. For example, regarding the massacre at the Little Big
Horn on June 22, 1876, and its possible effect on the nation,
he wrote in July to his friend, young Thomas Roberts:

> The death of Gen. Custer has created some feeling in two direc-
> tions: on the part of old settlers who were here in '62' in the Sioux
> massacre, a cry for extermination or at least a turn over to War
> Dept. Among others (the enthusiastic-peace people and those not
> personally acquainted with Indian war) a deprecation of the
> bloodshed which is likely to ensue this trouble. I am inclined [to]
> sympathize with the latter class.

The degree, bachelor of science, was conferred on him in
June, 1880, four years and two terms after he entered the
University of Minnesota. The local newspaper covered commencement and printed a story in which they mentioned that
"Mr. Clarence L. Herrick of Minneapolis recited briefly but
succinctly the dream or fable of Undine, and ably brought out a
hidden truth and applied it."[9]

Clarence Herrick had completed his college education, gained
teaching experience, served an apprenticeship as a field naturalist,
and published five articles in scientific journals. He needed a job,
and Winchell asked him to join the Geological and Natural History
Survey of Minnesota full time. The university had given him a

part-time instructorship in Botany during his last year and was pleased to have him register for the master of science degree. Clarence welcomed these opportunities. No offer would have been accepted with more enthusiasm than that from Director Winchell for continuing to work on the Survey.

•

NOTES

1. Paul Bert was a French physiologist, brilliant pupil of Claude Bernard.
2. Three essays by Charles Judson Herrick were published in Volume 54 of the *Scientific Monthly*. These were drawn from an unpublished book manuscript.
3. The instrument pictured was given to my father when he was a boy. It may be smaller than that of Herrick but has essentially the same mechanism, similar to that of Benjamin Martin's microscope of *c*. 1740.
4. C. J. Herrick, 1955.
5. J. Gray, *The University of Minnesota 1851-1951* (Minneapolis: University of Minnesota Press, 1956).
6. The Library of the University of Minnesota subscribed to the following scientific periodicals: *American Journal of Science and Arts; American Naturalist; Annales de Chimie; Microscopic Journal; Nature; Popular Science Monthly;* and *Scientific American* and *Supplement*.
7. Winchell had studied at the University of Michigan where he received the A.M. degree in 1869—the most advanced earned degree conferred by American universities at that time. Soon after his appointment on the Minnesota faculty he was relieved of much of the teaching so that he could devote more time to the Survey. He gave up his professorship in 1877.
8. The University of Minnesota refuses to release transcripts of grades without "permission from the student or his heirs." I

could locate no heirs, but Dr. Thomas A. Roberts, director
of the Minnesota Museum of Natural History, had obtained
this information in 1939.

9. It is difficult to connect the fable of a female water sprite
marrying a mortal to receive a human soul, with Clarence's
interest in natural history, but it does perhaps foretell his
concern with the body-soul concept of the times.

IV

POSTGRADUATE YEARS

Clarence Herrick aimed to continue his education, but he had to find means of support. His participation in work at home declined after he graduated from the university in June of 1880. His younger brothers assumed responsibility for the chores, which had increased because of their father's ill health, and this enabled Clarence to devote all his time to earning and saving money for graduate education. With the backing of Director Winchell, he became a full-time member of the Geological and Natural History Survey of Minnesota and later was promoted to the post of State Mammologist.[1] His work with the Survey required him to be away from home on frequent field trips.

Herrick entered upon his duties as a naturalist with the same intensity that had characterized his previous endeavors. He even sought to enlarge the scope of his work, and at one point requested that he be permitted to embark on some experiments in embryology, but the director advised against it. His early concentration on collecting and studying crustacea resulted in publication of a monograph that was profusely illustrated with his drawings, many from observations through the microscope. This appeared as a final report of the Survey for the year 1883.[2] It was Herrick's first major published work, part of which probably served as his thesis for the master of science degree.

Herrick's artistic talents had been recognized for some time. Even as a schoolboy he found amusement in making drawings on his slate. The plates accompanying his article on the crustacea contain drawings of professional quality which greatly enhanced the value of his monograph. Herrick constantly sought means of improving reproduction of his drawings, and on one occasion in 1882 wrote to Thomas Roberts:

33

> I am trying to get hold of some convenient way of recording notes permanently in way of drawings upon stone to be printed directly. If successful I shall escape the difficulty in getting engravings done by the non-scientific workmen. The photo-engraving process by which my last articles are illustrated is the best thing I have yet found. . . . I will send you the first results of the experiment.

In this same letter he said that he would like also to try "microphotography." He enclosed a proof sheet of a lithograph showing parts of some crustacea.

Soon thereafter, he began work on a laboratory manual for students of zoology, illustrating it profusely with lithographs. A photograph of one of his engraved stones and a copy of the lithograph made from one half of the stone are shown opposite page 65. The manual was privately printed in 1883.

No classroom attendance was required for the master's degree at the University of Minnesota in the 1880s. A candidate, to qualify, must have obtained a bachelor's degree at that or another institution and must have been engaged in scientific work for two years. An examination was then given on two branches of natural or physical sciences and three other subjects, such as mathematics, philosophy, and ancient or modern languages.

The master's was the highest academic degree offered. None of the university faculty of seventeen professors and instructors, including President Folwell himself, held the doctorate. Clarence Herrick expected to receive his master's degree in 1882 and later gave that as the date in his biographical sketch in the first volume of *Who's Who in America,* but for some reason the degree was not conferred until June, 1885.

He gave no serious thought to earning the Ph.D. degree, few of which were given by American universities at that time. Some colleges conferred them as honorary degrees or simply for presentation of a learned dissertation.[3] The University of Minnesota did confer the Ph.D. on Herrick *in absentia* in October, 1898, after he had become president of the University of New Mexico. He presented a thesis entitled "A Theory of Somatic Equilibrium with Illustrations of a Possible Mechanism Therefor in the Skin," which was based on articles published in the eighth volume of his *Journal of Comparative Neurology.*

Postgraduate study in Europe was the goal of many young Americans, few of whom could afford the luxury, travel grants being practically unknown. Herrick aspired to visit Germany and saved most of the money earned during his first year of full employment. He made it last as long as possible, but it gave him only one year.

He took leave from the Survey and went to New York on or about his twenty-third birthday, boarding a transatlantic steamer soon thereafter. His destination was Leipzig, where he enrolled as a graduate student in zoology under Rudolf Leukarts. He arrived about the first of July, but it was four months before he began to relate his experiences in letters to Thomas Roberts, who had gone to Philadelphia to study medicine. No earlier letters to family or friends survive.

The first one to Roberts was dated November 11, 1881:

I have been remarkably fortunate in my (for me questionable) undertaking. Everything has turned out very favorably. I left home with no bright anticipation and no rosy views of the future. . . . I made the journey quite comfortably and exceedingly cheaply and was so lucky, unaccustomed to travel as I am, to make no false moves. I came into a country and city where English is rarely spoken and with no adequate preparation was able not only to find my way and secure comfortable and cheap accommodations but also to acquire unexpected facilities for work— which was the principal thing. I have taken a foot tour to Dresden and seen some of Berlin and got an idea of the manners of country people in Germany and am content to do no more sightseeing till on my return. . . . I found no difficulty in getting work enough after once fairly hunting down the Zoological department of the University in this crazy old town, for you must know the University lies round loose all over Leipzig and every building that is not a church or the late residence of Goethe or Schiller is liable to be "Universitate Gebaude" and every body is either a student, a soldier, or a policeman unless he has a fiddle under his arm when, I take it, he is nondescript. The formalities of matriculation once over I found that my spare time, which before was all the time, was reduced to zero. . . .

I am at work just now hunting up the missing details in the blood circulation of some freshwater crustaceans (Daphnia) which have furnished material for as many monographs perhaps as any genus and are very well suited for microscopic study by their transparence. . . . The fun of following the blood corpuscles of an animal of this size [here he made a drawing 3 × 6 mm] through the branchial cavities in this living "critter" and studying the valves of a heart that a cambric needle would entirely obliterate and yet is beating 200 times a minute! is something extraordinary. Yet by two or three weeks study I hope to complete the circuit.

Later when his money was running low, he tried his hand in reporting for the hometown newspaper. One of his letters was published in the *Saint Paul Pioneer Press* with the dateline February 8. It was a gossip column of about 1,600 words entitled "From a Leipzig Student's Diary." Events described were a visit by the king of Saxony on a hunting excursion, music in the town, an amusing though disapproving account, "Prof. Delitsch and His Bottle," and finally some comments on education. This appears to have been the only one of his communications published by the *Saint Paul Pioneer Press,* and he was not very happy about it. He told his friend Roberts: "My articles for the *Press* . . . have been a source of regret to me—written at random times and some of them sent off with scarcely a second reading they must be very crude. My poverty has been my excuse and yet I do not get rich very fast thereby."

There is another letter to Roberts, written on February 19, 1882. Its preservation is most fortunate, for it gives a clue to Herrick's introduction to the works of Lotze:

It is now my plan to return in early summer for I feel that my object is so far accomplished that unless I decide to strive for degree it would hardly pay in proportion to the expense of time and money to remain. It seems entirely probable that by waiting until Fall I could get the "Phool D," but it would oblige me to take studies I could pursue to greater advantage at home. . . . I have also for a short time daily met the daughter of Sir Wm. Hamilton whose rich fund of varied learning has given me a favorable idea of the capacity of the English gentlewoman. Miss Hamilton was engaged in the translation into English of one of the works (*Mikrokosmos*, I believe) of Professor Lotze who was at the time of his death the leading German philosopher. Sickness interupted her work and deprived us of her company at meals.[4]

Was that how Herrick became acquainted with Hermann Lotze? If so, it was Elizabeth Hamilton who introduced him to the writings of the great German philosopher. A simple booklet by Lotze (possibly his last publication) had appeared in 1881 with the title *Grundzüge der Psychologie.*[5] This was a synopsis or outline of some of Lotze's lectures. Herrick purchased a copy, the tattered title page of which is shown on page 38. That marked the point at which his thoughts began to turn from the circulation of blood of *Daphnia* to the brain and mind (soul) of man. There is no good reason to believe that they dwelt seriously on neurology and psychology prior to that time, for none of his papers had been devoted to those subjects. Reorientation of his interests appears to have been inspired by his dinner-table conversations with Miss Hamilton.

Emulating the efforts of his friend, he began to translate Lotze's *Grundzüge* into English with the intention of having the booklet published in America. But no publisher could be found and he put the manuscript aside until he saw a need for a brief text in courses he was about to teach. Finally, and most significantly, he added a section on anatomy of the brain and had the book printed in Minneapolis early in 1885 at his own expense. He told his friend Roberts on April 28 of that year that "I am still doing some literary work and have just published a work on Psychology, a translation of Lotze."

The preface of his little book reads:

The translation of Lotze's *Grundzuege der Psychologie*, which forms the principal part of this book, was made in 1882, and it was expected to publish it at once. Circumstances prevented, for the time, the carrying out of the plan, but lately a personal need of some small text book to use in connection with the study of comparative anatomy in the scientific department of an under-graduate course led to its revival. This circumstance explains why the short chapter on anatomy has been appended unnecessarily as it may seem. It is believed that, in its present form, this volume will prove convenient; firstly for use in connection with the little that can usually be said upon the physiology of the nervous system in the comparative anatomy of our ordinary colleges, at the same time furnishing a thoroughly reliable foundation upon which to add the more extended work in psychology; secondly, as a first book of psychology where for any reason the physiological side does not receive special attention in the philosophical department. Of the value of Lotze's little work nothing need be

Grundzüge

der

Psychologie

Dictate aus den Vorlesungen

von

Hermann Lotze

Leipzig

Verlag von S. Hirzel

1881

said here, the great German philosopher is rapidly gaining recognition even in America, while the series of "Outlines," of which this forms a part, has been exceedingly well received abroad. Attention may be asked, however, to the fact that these are but *outlines,* and embrace but the dictated portions of an extended lecture course. Their use in the school room implies oral explanation and illustration, or, better, they may form the frame work for a lecture course which may deal as fully with anatomical and physiological details as time permits. It is a matter of regret that no English work in this department has as yet appeared which is not devoted to the indirect inculcation of a theory (unorthodox or otherwise), or for some other reason unadapted to place in the hands of college students.

The booklet is composed of three parts, the first two being the translations from the German text and the third, Herrick's new one on the brain.

Part I, designated "Individual Elements of the Inner Life," is physiological inasmuch as it covers such topics as sensation, nervous action, perception, memory, consciousness, and reflex action; the terminology is that of a philosopher rather than a physiologist.

Part II, called "The Soul," is more esoteric and in it one finds discussions of such topics as existence of the soul, reciprocal actions between soul and body, seat of the soul, realm of the soul, will, etc. The section on body-soul concept, in which Lotze concluded "the notion, not of the immaterial, but of the material requires to be demonstrated, and the gulf which seemed to separate body and soul as two completely heterogenous elements and thus to prevent their reciprocal action really does not exist," brought from Herrick the following footnote comments:

The reader will be interested to compare ideas briefly set forth in this section with the dicta of modern materialism. . . . Some of the various phases of materialistic thought on the relations of soul and body will be gathered from the following sentences: "Will is the necessary expression of a condition of the brain occasioned by external influences" (Moleschott). "Man is but the sum of parents and nurse, of place and time, of air and weather, of light and sound, and of food and clothing"; or, according to Feuerbach, "Der Mensch nur ist was er isst," *i.e.,* Man is but what he eats.

The final ultimatum—"Thought is as much a secretion of the brain as bile is of the liver or urine of the kidney" (C. Vogt) stands in bold contrast to the teaching of our author here and elsewhere.

Part III, on structure of the brain, was added in 1884 or early in 1885. By that time Herrick had become convinced that psychology, which he was being called upon to teach, must embrace structure of the brain as well as its function. He introduced the section "Primitive Elements" as follows:

> The animal body, complex and mysterious as it seems, and diverse and complicated as its various organs really are, is composed of nothing but cells and cell derivatives.
> As stated by Lotze, it is impossible to avoid conceiving of man as a unit-entity surrounded by or resident in a heterogenous agglomeration of corporeal units, and yet it is impossible to picture to ourselves the way in which the various processes which bring about sensations and which are really only states in various unlike bodily elements transmit the perfected product to our spiritual apprehension. However, although there is not physical analogy for the process it must be accepted as a fact, and we may, at least, be interested to learn just what adjustments in the body are necessary to such and such spiritual affectations, and through what mediation the spirit controls the body.
> The nervous system is no exception to the above statement, and all of that wonderful mechanism which we call the brain, and which is the physical basis of character and the soil in which the soul, in its corporeal relations, is rooted, is a mass of cells and cell products like the muscles and bones which do its bidding.

Herrick's account of the brain and nerves was treated developmentally and contained bits on neurochemistry as well as neurophysiology. It was exceedingly brief—twenty-six pages including seventeen elementary figures, conforming thus with the plan of Lotze's parts of the booklet. One would like to think that Herrick kept his account brief for that reason rather than because his own knowledge of the subject at that time was in a primitive state, but there is evidence favoring the latter view. Better accounts of nervous structure and function were available in 1885.[6] Herrick's early trend toward comparative neurology is revealed in the illustrations.

The publication of Herrick's translation of Lotze's *Grundzüge*

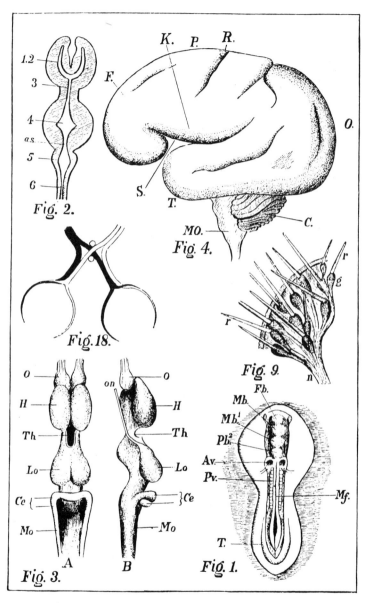

Reproduction of Plate I in Herrick's translation of Lotze's book, published in 1885.

was a landmark on the way toward shaping psychology into a true science. Incorporating an introduction on brain structure and function began its removal from the realm of philosophy to that of neuroscience. It is noteworthy that the little book has been re-printed by Arno Press (1973) in a series entitled *Classics of Psychology.* His discovery of Lotze's writings and translation of the *Outlines* in 1882 was without a doubt Herrick's most significant accomplishment during his year in Germany.

Clarence Herrick returned from Europe in the summer of 1882. Thoughts of the brain and mind were put aside as he resumed his duties with the Geological and Natural History Survey of Minnesota. He had to complete his collections of mammals, and many drawings were needed to illustrate the monograph he was expected to submit to the director. Nonetheless, when winter approached he again interrupted his work.

His father's precarious health led to a decision to travel south and spend some months in Alabama, where he hoped to find a warm climate and to visit people he had met in the Reconstruction days following the Civil War. He could not travel alone, so Clarence accompanied him and, true to character, took along a microscope and spent his time collecting, identifying, and drawing minute specimens from the pools and ditches. The two men returned to Minneapolis at the end of the year with no significant improvement in the constitution of the senior Herrick, who lived on until the spring of 1886.

The decision to return in the dead of winter may have been influenced by Clarence's desire to be with his fiancée, Alice Keith, who was teaching in a Minneapolis school. They were married in June of 1883 and spent a honeymoon camping on the St. Louis River in northern Minnesota where Clarence continued his study of mammals for the Survey which was to be completed two years later.

An invitation came from the president of Denison University to go to Granville, Ohio, as a substitute for the professor of Geology and Natural History, who was in England. Herrick obtained another leave of absence from the Geological and Natural History Survey of Minnesota and accepted the temporary instructorship for the autumn term of 1884. After three exciting months at Denison University, he returned to Minneapolis with a reasonable assurance that his future held a professorship.

But first it was essential to complete the text and illustrations of a final report on mammals. He did so during the ensuing year and submitted a long manuscript filled with artistic drawings, many in color rivaling those of Audubon. He expected this to be published by the state of Minnesota, but unfortunately the legislature would not provide the rather substantial sum of money required. Seven years later, parts of the monograph and a few of the illustrations were printed as *Bulletin 7 of the Geological and Natural History Survey of Minnesota.* The reluctance to finance the monumental original work may have been influenced to some extent by Herrick's resignation from his position as State Mammologist to join the faculty of Denison University.

NOTES

1. Director Winchell stated in his report for 1883 that "Mr. C. L. Herrick has been given the mammology of the state, with a view to the collection of skins and skeletons for the museum, and the preparation of a final report on the same for publication in about two years." Herrick completed the task on time.

2. C. L. Herrick, A final report on the Crustacea of Minnesota including the orders *Cladocera* and *Copepoda. Twelfth Annual Report of the Geological and Natural History Survey of Minnesota for 1883.* Part V, pp. 1-192, 30 pls., 1884.

3. The first edition of *American Men of Science,* published twenty-four years later, contained the names of nearly 4,000 men and women, only 46 percent of whom had the Ph.D. or D.Sc., and more than one-fifth of these degrees had been earned abroad. Only forty-nine of the scientists living in 1906 had obtained the doctorate prior to 1880.

4. She died in 1882, possibly as the result of the illness mentioned by Herrick. Elizabeth Hamilton was the daughter of

Sir William Hamilton, Bart. (1788-1856), distinguished Scottish metaphysician and scholar of German philosophy. Upon Elizabeth's death the translation of Lotze's *Mikrokosmus* was completed by Miss Constance Jones of Girton College, Cambridge, and published in New York by Scribner & Welford in 1886.

5. H. Lotze, *Grundzüge der Psychologie. Dictate aus den Vorlesungen* (Leipzig: G. Hirzel, 1881). Rudolf Hermann Lotze, born in 1817, studied medicine at the University of Leipzig where he became interested in physiology and psychology and came under the influence of Weber and Fechner. Nominally a philosopher, he lectured on psychology for thirty-seven years at the University of Gottingen. He developed a view of psychology that was to influence many future scientists. Of special note were his textbook in physiological psychology (the first), his views of therapy and psychopathology, and his theory of spatial perception based upon the concept of "local sign." His lecture notes were published after his death in 1881.

6. The popular book *The Human Body* contains accounts of neural structure and function that were authoritative at that time. The book was written by H. Newell Martin, a former pupil of the noted English physiologist Michael Foster, and published by Henry Holt and Company of New York in 1881. It had gone through six editions by 1890. One could purchase it either with or without a section on anatomy and physiology of reproduction. In the complete version, sex was relegated to an appendix, separately paginated and separately indexed! H. Newell Martin, D.Sc., M.A., M.D., F.R.S., was Professor of Biology in the Johns Hopkins University; Fellow of University College, London; late Fellow of Christ College, Cambridge. This book may have been unknown to Herrick in 1885.

V

LAUNCHING AN
ACADEMIC CAREER

Clarence Luther Herrick's first permanent academic appointment, the Chair of Geology and Natural History in Denison University, came a few months before his twenty-seventh birthday. His qualifications for a professorship were impressive: a year of graduate study at the University of Leipzig where he had acquired fluency in the German language; a master of science degree from his alma mater; instructorships at Minnesota and Denison; extensive experience as a field naturalist with the Geology and Natural History Survey of Minnesota; and even a brief and rugged term as schoolmaster in a Minneapolis district school. Moreover, he had sixteen publications, including a small book. Herrick had an outgoing personality and a commanding presence on the lecture platform where with ambidextrous blackboard sketching he displayed a fine sense of showmanship. But there was another qualification that, above all others, led to his being selected: He was an earnest Christian—a Baptist.

A faculty vacancy had occurred at Denison University, and the following report of a search committee was adopted by the trustees on June 23, 1885:

> Your Committee to whom was referred the question of filling the vacancy occasioned by the resignation of Professor Hicks have given the matter careful and prayerful consideration. Their attention was very early called to Mr. C. L. Herrick of Minneapolis. Correspondence and inquiry developed the fact that although still a young man he had gained an excellent standing in his chosen department. Since his graduation in 1880 he has spent some time in Germany and afterward was employed in the University of Minnesota, partly as instructor and partly in the State Geological Survey. He has done much original work as an investigator, some of which has been given to the public. From his pastor and

45

Mammals of Minnesota.
—×-×—
C. L. HERRICK.
2450 Pleasant Ave.

Minneapolis, July 16ᵗʰ 1885

Dear Bro.,

I am today in receipt of your letter of June 30ᵗʰ informing me of my election to the chair of Geology and Natural History in Denison University. In accepting, as I now do, the trust offered me I beg to express a high appreciation of the honor done me by your honorable body and of the responsibility involved in its acceptance, and trusting that the future may in some measure justify the confidence thus expressed.

I remain

Very respectfully yours

C. L. Herrick

W. C. P. Rhoades D.D.
Sec. Board of Trustees Denison University.

others we learn that he is deeply and actively interested in church work. We found him under an engagement in Minnesota, but one from which he could obtain leave of absence and accordingly we secured his services for work of the Fall term. Personal acquaintance and observation of his work confirmed the impression already made and we are hearty and unanimous in recommending his election to this Professorship; and we may add that we regard the way in which we have been led in this matter as another indication of the care of Divine Providence for this Institution.[1]

The secretary of the Board of Trustees, W. C. P. Rhodes, D.D., promptly dispatched the letter of invitation. On the tenth of July, Herrick answered; his letter is reproduced opposite.

The professorship carried a stipend of $1,700 per year, one hundred dollars more than he had expected and quite adequate for his needs. He could bring his wife, Alice, and their seven-month-old son, Henry Nathan Herrick II, to Granville, which he did in time to start the autumn quarter of 1885.

Denison University had been founded at Granville, Ohio, in 1831 by Baptist ministers as the Granville Literary and Theological Institute and was a very small college when Herrick joined its faculty. The college students numbered scarcely more than fifty, and there was a slightly greater number of pupils in the preparatory department. There were nine faculty members, including the president who was the only administrative officer.

Most American institutions of learning at that time were experiencing the impact of science upon their curricula and faculties, and Denison was no exception. Two full professorships in science had been established and the bachelor of science degree was being conferred after a four-year course of study. The number of students preparing for the ministry was diminishing. The new biology was in its ascendency at Denison as elsewhere and a conflict between it and religion smoldered. A teacher had to tread lightly over several issues, especially evolution of man and the body-soul concept.

Denison University had lost both of its teachers of scientific subjects in 1884. The professor of Chemistry and Physics had died and was succeeded by Alfred Cole. The professor of Geology and Natural History, Leslie Hicks, who had been studying at the

British Museum on a year's leave of absence, decided not to return and his resignation was accepted on June 24, 1884.[2] Herrick had substituted for him that autumn with a stipend of $800 and became his permanent replacement in 1885. The two new professors were to have major impact upon science at Denison University.

Herrick's first impression of the university had pleased him. He wrote to Thomas Roberts in October, 1884:

> The college is second in endowment in the State, I believe, and has maintained a good rank of scholarship in the classical and mathematical departments. The scientific dept. has been made the scapegoat, but this is now to be remedied by the formation of an English course for those of inferior advantages or ability. Funds are short now but are promised. There is a need of illustrative material and of books greater than that at the U. of Minn. even, in some lines amounting to entire absence. This difficulty must be overcome & this will require work. However, I ordered a number of Baush & Lomb "model" microscopes ($50) & we are now about finishing a course of advanced botany in which the students have spent about an equal amount of time in laboratory & lecture work. . . . The offer is now made definitely accompanied by a salary of $1,600 for next year with farther increase in the following year. I have not yet committed myself but cannot help feel that my duty lies in this direction. . . . It will mean much less of personal original work yet even in that point of view it may be that 5 years of this sort of thing will be a good preparation for some other work and by that time maybe some young fellow will have grown up here ready to carry some of the burden.

It appears that Herrick had not reached a firm decision to go ahead with a teaching career which would limit his time for research. But in spite of lingering doubts, he had had the possibility in mind for several years. The usefulness of his translation of Lotze's outline for elementary courses in psychology had occurred to him. Moreover, he had published the first part of a laboratory manual for Zoology entitled *Types of Animal Life Selected for Laboratory Use in Inland Districts. Part I— Arthropodia.*[3] In the introduction of that booklet he expressed his belief that minute transparent freshwater animals could be used more advantageously in student laboratories than preserved saltwater specimens that required dissection.

Herrick was an effective teacher and during his 1884 term won the admiration of colleagues on the Denison faculty as well as his students. Geology was the subject of his major responsibility and, as it required no extensive laboratory, he taught the classes in the field. He took his students on walks, discussing things they encountered along the way and discoursing on philosophical subjects. He did not conduct the usual recitations of memorized textbook material. His was a different kind of pedagogy, and those who went with him into the field that first term at Denison were stimulated by the experience and charmed by the young teacher. One of his students, H. H. Bawden, years later wrote about his qualities as a teacher:

> His phenomenal success as a teacher . . . was due to factors some of which were easily seen—others are harder to define. After his attractive personal qualities and magnetic enthusiasm, I should place his deep philosophic insight and the fearless way in which he disclosed his profound thinking to the least initiated of his pupils. The ability to do this without befogging the air was an exceedingly rare gift and was stimulating even to a dullard.

Herrick's activities were not confined to teaching but included participation in extracurricular affairs. For example, he became a member of the executive committee of the Association of Colleges of Ohio, his term expiring in 1888.

During the first "probationary" period in Granville, one of the second-year students, August Foerste (who later became well known in geology), introduced his young instructor to fossil-collecting sites of the county. The latter's interest was immediately aroused and he lost no time in beginning a study of rock formations and their fossils, an activity that claimed much of his time over the next four years, during which he published a series of articles on the geology of Licking County, Ohio, that established his reputation as a leader in the field.

It was on one of their collecting excursions that Herrick and

Foerste discussed advantages of having a scientific magazine of their own. The idea appealed to both of them and they set about preparing articles that might be published in such a journal, the launching of which became Herrick's first order of business when he returned the following year.

Herrick went back to his family in Minneapolis at the end of the 1884 term and resumed his work on the Geological and Natural History Survey of Minnesota. He completed the drawings and text of the monograph on mammals of Minnesota while waiting for official word of his appointment from Granville.

Immediately after settling in Granville in the autumn of 1885, he and Foerste renewed plans for their scientific periodical. Herrick named this the *Bulletin of the Scientific Laboratories of Denison University*.[4] Its first issue appeared in December and continued quarterly. The *Bulletin* was financed by money from several sources. The university trustees provided a small subsidy, Herrick and Foerste each put in a little money, and various friends contributed some. The printing was done locally, but the editor helped the printer execute the lithographs for which Herrick and Foerste had made the stone engravings in their laboratory. The title page is reproduced opposite.

There were six original articles in the first issue of the *Bulletin,* two by Foerste and four by Herrick, on ornithological, botanical, zoological, and geological topics, but none on neuroscience. An artistically rendered, hand-colored lithograph of the evening grosbeak *(Hesperiphona vespertina, Bonap.)* served as a frontispiece. The articles were profusely illustrated; the execution of drawings appeared to command greater care than composition of text, which in some cases seemed simply to support the pictures.

Herrick appeared to be well pleased with his life in Granville. On the last day of January, 1886, he wrote to Thomas Roberts, of Minneapolis:

> The quiet Sunday ever affords me time to write a line. I do not hear a word directly or indirectly of your doings or beings, and hope that the silence is not ominous. We are thoroughly shut out of our old world here and I am too busy to keep up correspondence. In fact, I am busier than ever before. But all passes pleasantly, and success in fair measure is met. . . . Our class is study-

BULLETIN

SCIENTIFIC LABORATORIES

OF

DENISON UNIVERSITY,

EDITED BY

C. L. HERRICK, M. S.,

PROFESSOR OF GEOLOGY AND NATURAL HISTORY.

VOL. I.

GRANVILLE, OHIO, DEC., 1885.

ing histology in a small way—hardening and making sections of liver, kidney, stomach, etc., with fair results. We have Bausch & Lomb's new Microtome which is a "beauty"—much superior to those in use in Leipzig. We are also doing lithology[5] this term & these two new studies keep me at work. . . . We seem to be making considerable progress, although considerable apathy is to be expected.

It is surprising that he had nothing to say to his old colleague of the Young Naturalists' Society about his new periodical, the first issue of which had appeared only a little more than a month previously.

When the first excitement of the professorship subsided and an awareness of many deficiencies at the university began to arise, Herrick could see some of the serious obstacles he would have to overcome. The two most obvious ones were lack of a scientific library and inadequacy of laboratory space and equipment for experiments. The college advertised in 1885 that it had "eleven Professors and Instructors [including the president], well-equipped Chemical, Physical and Biological laboratories and a large and excellent library [note the lower case *l*]." The last claim was patently an exaggeration, true only in regard to the appearance of the building.

Herrick's reasons for initiating the publication of the *Bulletin* were not clearly stated. An editorial simply declared that "every well conducted institution of learning should form a recognized centre of scientific activity; and legitimately concerns itself, not only with the instruction of those who directly entrust themselves to its charge, but with the dissemination and conservation of information relating to the subjects taught." However, there was a more fundamental reason for having a scientific periodical: the need to acquire world literature by exchanging the *Bulletin* for publications of foreign academies and other institutions. Herrick had seen that done successfully at Minnesota.

Granville, a village of some 1,200 people, had a public library, but the books (carefully censored) were of no value to students of science. The university had built an attractive library building in 1878, but it contained so little of value that a library committee in 1891 reported its contents to be mostly trash. The president and faculty pleaded for money with which to buy books so that their students, isolated as they were, could know what wonderful things were happening in the world. But the trustees turned deaf ears, apparently in the conviction that the word of God was quite sufficient, and besides, the brethren had many other pressing needs for funds.[6] This appalling situation faced Herrick, who tried to remedy it in some small part by establishing a periodical which he could exchange.

Seemingly as an afterthought, Herrick organized a scientific society which met first on April 16, 1887, as the Denison Scientific Association, with twenty-seven charter members most of whom were students (there were only eleven faculty members of all disciplines). The *Bulletin* thereafter became the organ of the association, but there was an alternative plan. The first and only issue of *Memoirs of the Denison Scientific Association*[7] appeared that year. Although the association prospered and has continued to meet since its founding, its *Memoirs* died with birth.

Herrick's article in the *Memoirs* described crustacea he had collected several years previously in Alabama. As the school year wore on and frustrations continued through a cold winter, he began to recall warm days on the Gulf of Mexico. He imparted a wishful dream to his friend Roberts in a letter dated March 13, 1887: "It is proposed to form in conjunction with our own laboratory a sea-side summer laboratory on the Gulf. I send invitation in advance to spend a summer with us on the semi-salted sea. . . ." But he seemed to have had second thoughts when he recalled a bout of malaria, writing again in October to his physician friend:

> You will [be] professionally interested perhaps to learn that no return of the malaria has occurred. . . . I therefore announce myself no longer a patient and return to my disbelief in medicine (when not sick). . . . We are progressing rapidly here. A new scientific building and museum are to be soon put up at the expense of $25-30,000 which will enable me to put in existence certain schemes entertained as long ago as when a member of the Young Naturalists' Club.

This was a preview of Herrick's next project.

Laboratories for natural history were at best very primitive.

Herrick quickly came to realize that they were inadequate for the experiments he had hoped to do when he returned from Germany. Nevertheless, he and a student, William G. Tight, were conducting experiments on the cerebral cortex of several species of animal, stimulating various regions and subsequently examining histological differences between motor and sensory regions. They needed space.

President Anderson was aware of the deficiencies. He recommended to the Board of Trustees, meeting in June, 1887, that "*Special Scientific Courses* should be founded" and noted that "the *Department of Geology and Natural History* needs laboratory and apparatus." He informed the Board that "Prof. Herrick a profound scholar and a prodigious worker . . . is bringing the University into connection with the best colleges and universities of our own country and with some of the universities of Europe." The trustees authorized purchase of one microscope costing $50.

According to a notice in the *Bulletin of the Scientific Laboratories,* the Natural History Department offered work in physiology of the nervous system and comparative psychology, but their facilities for the laboratory work were limited. This was implied in the following quotation:

> In the sophomore year the winter term is devoted to Comparative and Human Physiology and Hygene. The genesis of organs and comparative (vertebrate) morphology is discussed as far as time permits. The hygenic applications of physiology are briefly presented but physiology of the nervous system and comparative Psychology are relegated to the elective term of the Junior Year. An amount of time equivalent to an hour per week [!] is devoted to dissection and other laboratory practice. . . .

Herrick schemed to relieve the deficiency of laboratory facilities by construction. He obtained plans for a new science building from an architect in Newark, Ohio. It is not recorded that the college paid the architect, and I assume that Herrick did so himself. The plans were published in Volume IV of the *Bulletin* in 1888, and part of them are reproduced here.

No money was available for construction, but he believed that the plans might be used to elicit interest of prospective donors.

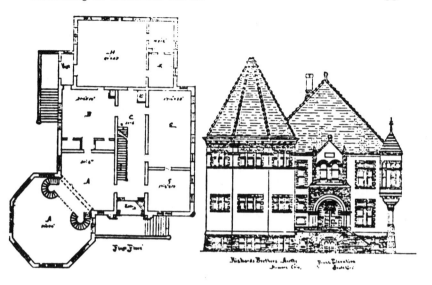

Architect's drawing for a hall of science that Herrick proposed for Denison University in 1888.

They probably influenced Eugene Barney to contribute the money (twice the amount originally contemplated by Herrick) for an even more commodious hall of science several years later.

Herrick apparently received some money from Minneapolis after his father died. He was able to buy two acres of land adjacent to the college and build a ten-room house on it in 1888; a photograph showing its present appearance is shown opposite page 64. He intended to build animal pens and breeding sheds, facilities that were sorely needed for experimental psychology, but no sooner had his family occupied their new house than he received an offer of a professorship at the University of Cincinnati.

Herrick had become restless by that time. He wrote to Thomas Roberts in the fall of 1887, saying: "I do not know how long I can stand the strain imposed by the position here. I hope to last three years more, after which if the way opens, special work in Biology is on the docket." The chance to move came two years before expected. His decision to do so had no basis in expectation of financial reward. It was an emotional response, and I can only speculate on the reasons behind it. He may have been impatient with the university administration. President Galusha Anderson was engaged in a strenuous effort to unite the women's and men's colleges and combine their facilities, and had little time to press forward Herrick's plans for a science hall that would provide for experimental psychology as well as other sciences; Cincinnati had a new building. Then, too, there was a smoldering conflict between theology and biological science, while a more relaxed atmosphere prevailed in Cincinnati. Whatever the reasons, to have quite suddenly vacated the new house and dropped his plans for a science building and an animal resource for his neurological research must have resulted from a combination of disturbing events. His mother's illness may have contributed,[8] but his own unstable emotional state appears to have been the principal one.

The Board of Trustees of Denison University met in June, 1888, and the president "recommended that leave of absence be granted to Prof. Herrick for recuperation for such time and/or such conditions as may relieve him from anxiety and give reasonable hope of retaining his services." The Executive Committee of the Trustees met on July 3, 1888, and moved "that Prof. Herrick receive $800.00 of his salary during the coming year of his absence from duty. Carried." His resignation followed a year later.

NOTES

1. Signed by Alfred Owen, D.D., the president of Denison University.
2. Professor Hicks had been the foremost champion of science at Denison during his tenure from 1870 to 1884. He strove to raise the status of scientific teaching, but usually was frustrated, on one occasion being forced to cancel four lectures on Darwinism. (Anti-Darwinism was still strong forty years later.) There is little doubt that Hicks resigned from the Denison faculty when the opportunity arose to go to a more liberal institution, one that placed a higher value on science.

3. Privately printed in Minneapolis, 1883.
4. It has been published continuously, although the name was changed in 1922 to *Journal of the Scientific Laboratories of Denison University.*
5. Lithology, the study of rocks.
6. The librarian's financial report rendered in 1889 covered Herrick's last year. It showed receipts: $77.77; disbursements: $69.20; balance: $10.57.
7. *Memoirs of the Denison Scientific Association* 1: 1-56, 7 pls., 1887.
8. Anna Herrick suffered a manicdepressive psychosis after her husband died in 1886 and she could not be left in Minneapolis. She was hospitalized in Rochester, Minnesota, and later discharged. She came to Granville to keep house for her sons William and Charles, and was there at the time Clarence moved to Cincinnati.

VI

CINCINNATI

Herrick may have expected the new environment to provide cultural advantages that he had not enjoyed since his year in Leipzig, Germany. Cincinnati, settled a century previously, had become a thriving city with a population of more than a quarter million people, among whom were many Germans who had brought with them a love of music that contributed importantly toward making the city a cultural center. Literary interests, too, were strong. The city boasted that its Literary Club was the oldest organization of its kind in America. Library facilities were good. The main public library even contained a large collection of medical books and periodicals, although publications in the biological sciences were inadequate to satisfy all of Herrick's requirements.

The University of Cincinnati had been established in 1873 by reorganization of an earlier Cincinnati College that dated back to 1818. It was a municipal, coeducational institution that was in a phase of growth and expansion, soon to lead to incorporation of schools of law, medicine, and dentistry. The first commencement of the university, held in 1878, had stirred considerable public interest. Four men and one woman received the bachelor of arts degree, and the bachelor of science was given to one man who submitted a thesis entitled "The Roots of Communism."* By 1888 the enrollment in the Academic Division (i.e., Liberal Arts) was nearly three times that at Denison University. Moreover, there was greater emphasis on science, especially in relation to medicine.

Herrick's new environment was more liberal than that in Granville. Some held it to be godless. Bible reading in the university

---

*Not in the present sense.

had been placed on a voluntary basis. Indeed, the civil courts had declared that enforced reading of the Scriptures in public schools was unconstitutional; and the University of Cincinnati was a public school.

Denison University granted him leave to go to Cincinnati, and it was not until the next spring (1889) that he decided to stay. The *McMicken Review* of the University of Cincinnati noted in its April issue: "Professor Herrick has formally accepted the Chair of Biology. This means, we take it, that he will stay with us."

The college classes were conducted in a handsome brick building (four stories and an attic) that was only about five years old. Herrick was assigned a large laboratory and classroom, off of which there was a small office in an alcove. He found little use for the office; the laboratory formed the center for most of his neuro-biological investigations.

He took up his Cincinnati post in September, meeting classes and getting his research organized, while keeping in close touch with colleagues in Granville. He set his students to work in small groups on individual projects, and once a week held an informal seminar at which tea and cakes were served. He gave no examinations. His lectures illustrated with bimanual blackboard drawings, were well received. In January, 1889, The *McMicken Review* reported that "Professor Herrick has completely captivated his students in the Biological Department. He bids fair to become one of the most popular instructors in the University."

Herrick laid his geological research on the shelf, as it were. With more adequate space and equipment for anatomical and physiological studies, he could begin to do those things that he had hoped to do after his year in Germany. The move to Cincinnati was accompanied by a burst of activity resulting in amazing productivity in neurobiology.

His first publication of original work on the nervous system appeared in the *Cincinnati Lancet-Clinic* one year after he arrived at the university. In it he went directly into the problem of cerebral localization, a field in which his interest had been aroused by the writings of Fritsch and Hitzig. He began his article: "The striking results of recent investigations upon the physiology of various parts of the hemispheres, conflicting as they are, have given a fresh impetus to the study of the minute structure of the cortex." He mentioned work of others, namely, Luciana and Seppilli, Luys, and Beven Lewis, but gave no references.[1]

His principal experimental animal was the groundhog, or woodchuck, but he drew comparisons with the brains of rabbit, raccoon, and opossum. He used electrical stimuli to explore the cortical surface in search of motor centers for forelimbs, hind limbs, face, and neck. Small, numbered pegs were inserted into the cortex to identify points for histological study, and diagrammatic maps of the cortex were constructed. He reported his technique: "The brains were placed in chrom-acetic solution twenty-four hours and then in alcohol, and continuous series of sections in various directions mounted in balsam. Several hundred such sections were prepared and studied by the method of geometric reconstruction from camera drawings and measurements."

Herrick tried to identify different types of nerve cell by appearances after staining in the chromate solution, noting location of cells of different size and variation in arrangement of dendrites, especially basal dendrites of large and small pyramidal cells. He did not make complete maps of the cortex but concentrated on an attempt to distinguish between motor and sensory elements. He discussed his findings in relation to reports of Munk and Meynert (no references) and in conclusion declared that "if we can demonstrate in the cortex afferent and efferent projection systems, and distinguish the cells occupied with psycho-motor and psycho-sensory processes, certainly a great step is taken toward an intelligible construction of cerebral mechanics." This is a concept of modern neuroscience.

Herrick's next report was presented on February 4, 1889, to the Cincinnati Society of Natural History by title and published in their *Journal* in 1890. It was an anatomical study on the brain of the alligator, the first in a series on comparative neurology of inframammalian species. Four small alligators had been given to him by colleagues, and he prepared their brains by histological methods similar to those used for the woodchuck brain. He used supplementary stains: in bulk with alum cochineal; in sections with hematoxylin or analine blue-black. This thirty-three-page article was illustrated with nine double-page plates of lithographs. A shorter paper on the cells of the cerebral cortex appeared in the *Microscope* the same year.

The cortical localization experiments, which had been started in Granville with William Tight, appeared with the title "The Central Nervous System of Rodents—Preliminary Paper" in 1890. This comparative neurological study of rodents included brains of the opossum and cat. Although the authors called their paper preliminary, it occupied sixty pages and had nineteen plates of figures. The physiological experiments were performed after inducing anesthesia with ether and chloroform. The cortex was stimulated with bipolar platinum electrodes, "Du Bois-Reymond coil and Genet cell," the current strength being tolerable to the tip of the tongue.

The authors did not include a bibliography but mentioned older works of Max Schultze, Baillarger, and Gerlock, and recent articles by Koelliker, Meynert, Arndt, Hitzig, Deiter, Stilling, Nanssen, and Fritsch. They noted that a paper by Golgi in *Revista Sperimentale* (1883) was unavailable to them. The sources of their knowledge of the literature probably included journals received in exchange for the *Bulletin of the Scientific Laboratories of Denison University*. By 1889 the *Bulletin* was exchanged for 104 periodicals of which 54 were foreign.

Herrick attended the meetings of the American Association for Advancement of Science in Philadelphia in 1890. Among his unpublished papers is the manuscript "Suggestions as to the Origin of the Neuroglia of the Vertebrate Brain," describing a study he had carried out with Golgi's method. Perhaps this was the paper given at those meetings.

Another unpublished manuscript credited to the University of Cincinnati bears the title: "A Case of Morbid Affection of the Eye in the Cat Accompanied by Degeneration of the Occipital Cortex of the Opposite Side." This case report represents a foray into the field of neuropathology.

The next year found Herrick busy with research and writing,

in addition to carrying on an active teaching program. He published two short articles, one in *Science,* "The Evolution of the Cerebellum," the other in the *Anatomische Anzeiger,* "The Commissures and Histology of the Teleost Brain." The latter he signed as a professor in the University of Chicago, but the work on which it was based was done in Cincinnati. Both these short articles were "pot boilers" intended to be incorporated into major publications.

Herrick joined the Association of American Anatomists in 1891, and his name appears with ninety-four other members in the proceedings of the fourth annual meeting. He gave his address as Granville, Ohio. He attended the fifth meeting, which was held in Princeton, New Jersey, December 27 to 29, 1892, and gave two reports, "Histology and Physiology of the Nervous Elements" and "Embryological Notes on the Brain of the Snake." Two years later he was named as delegate to the American Congress of Physicians and Surgeons for the association. His name was dropped from the list of members in 1897 after ill health had forced him to become inactive.

The most important event of 1891 was the appearance of the *Journal of Comparative Neurology,* the first issue of which bore a March date. The founding and early history of this periodical will be considered in the next chapter. The first volume contains a series of articles, based on Herrick's research in Cincinnati, all of them concernesd with comparative morphology of the central nervous system. An important contribution by his student Charles H. Turner, entitled "The Morphology of the Avian Brain," likewise appeared in this volume.

Although Herrick's work in Cincinnati came to a halt in December, 1891, articles based on his research there continued to appear in the *Journal of Comparative Neurology.* Morphology of the brain of primitive species developed into a major interest, but cortical localization was not neglected, and a short paper entitled "Localization in the Cat," based on experiments conducted with the assistance of one of his students, E. G. Stanley, was reported in the second volume. They removed a narrow strip of the left cortex of a "half-grown kitten . . . extending from the crucial sulcus to the limits of the middle external gyrus caudad, and including nearly the whole of that gyrus . . . entire thickness of cortex and most of the white matter was removed. The animal was observed for four or five weeks during which time nearly all symptoms disappeared." Herrick concluded: "We are confident . . . that many of the contradictory results of experiments [of others] are due to proliferating regeneration which supply the lost material in the case of young animals."

Herrick's brief sojourn in Cincinnati represents the peak of his career in neurobiology. In his urge to learn ever more about the mysterious workings of the mind, his explorations touched upon many aspects of neurology and he came to the realization that approaching the problem through ontogeny and phylogeny might be the most productive. He drove ahead on preparation of brains for histological studies. He had no technician, but a few of his students worked with him. He searched the scientific literature; he wrote and he drew his own illustrations. These activities occupied most of his day. Not even the negotiations with William Rainey Harper of the University of Chicago, which occurred throughout 1891, stemmed his furious research drive.

Throughout the three years in Cincinnati Herrick maintained communication with Granville through letters to his colleagues and former students of Denison University. The new president of that college was trying to increase the support of science departments. Herrick's brother Charles attended a banquet of the Cincinnati alumni of Denison and afterward, on March 22, 1891, wrote to his fiancée, Mary Elizabeth Talbot of Granville: "There followed a short speech by Dr. Purinton [the new president] in which he pressed particularly the need of the proposed scientific building and new chair of Natural Science." Clarence Luther Herrick three years earlier had initiated the movement to obtain a hall of science.

Herrick had little time for activities other than those in the laboratory and libraries. I believe the rest of Cincinnati held fewer attractions than he had expected. Perhaps he hoped to find it a western Leipzig, for the city had a large population of German people. But Cincinnati was a river town with an even greater number of citizens who lacked culture—a rough lot for the most part, whose interest in the German breweries surpassed that in the German musicals. Herrick found few who reminded him of the gentle folk of Leipzig. His encounters with some of the townspeople annoyed him to the point of declaring that "one cannot live in the same world with people such as make up the bulk of Cincinnati." Moreover, he felt a lack of Christian fellowship that had been such an important part of his life in Granville and Minneapolis.

Herrick's work in the University of Cincinnati was greatly appreciated, and he was not at all unhappy within its walls. The university was rapidly becoming the kind of "federation of professional colleges" that had been envisioned by President Folwell at Minneapolis. There were opportunities in Cincinnati, such as the library facilities, that could be found in few other places in the Middle West. Only an outstanding opportunity elsewhere could attract Herrick to leave, but that is what came about when Harper offered him a professorship at the new University of Chicago.

Cincinnati wanted to meet the Chicago offer, but Herrick declined it. He, of course, could not foresee his error. He had been at Cincinnati a little over three years when his resignation became effective at the end of December, 1891. The Board of Directors accepted it with regret, and entered into the minutes of their meeting a testimonial of their appreciation of his service to the university. When the chair he had occupied became vacant in 1894, they invited him to return, but by that time his health had failed and he could not accept it.

NOTE

1. The lack of specific references in Herrick's articles embarrassed his brother when the latter was preparing the biography (1955).

The house that Herrick built in Granville
in 1888. This photograph was taken in
1975.

Barney Science Hall at Denison
University, 1894. The building was
gutted by fire in 1905.

A lithographic stone engraved by Herrick about 1883 for illustrations in a privately printed book.

A lithograph from the right half of the stone.

A microscope similar to the one given to Herrick by his father in 1875.

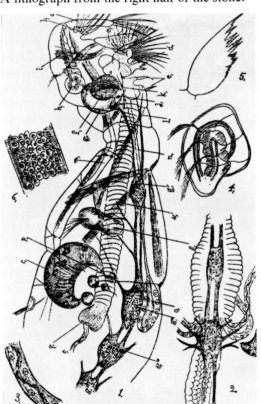

VII

JOURNAL OF
COMPARATIVE NEUROLOGY

Charles Judson Herrick told about the founding of the *Journal of Comparative Neurology* in an essay commemorating its fiftieth anniversary.[1] Several other neuroscientists who had been associated with the *Journal* during its early years likewise recalled events.[2] There is relatively complete knowledge about most of the seventy-eight years of this illustrious publication, but a few questions regarding the first years remain unanswered.

Why did Clarence Luther Herrick start the *Journal* when he did? There was no critical need for a periodical of its nature in the spring of 1891. His personal research program, he recognized, could be benefited as it had been when he successfully founded the *Bulletin of the Scientific Laboratories of Denison University* six years previously. That had provided a means of exchange for some of the world's best publications in natural history, which he sorely needed. Would it not be worthwhile to effect exchanges with publishers of other periodicals, particularly those in which neurological and psychological articles appeared? Little of note was being published in America, but a good many reports were coming out in foreign journals. For that purpose he required a more specialized bulletin than the one at Denison University. It is evident that this was one of Herrick's considerations in founding the *Journal of Comparative Neurology,* for he wrote from Berlin to his brother in Granville, requesting twenty copies of each issue of Volume I that he could use for establishing exchanges and for advertising.

Charles Judson Herrick, who was a student in his brother's department and was living in his household at the time, was given

no confidences and knew nothing about the plans and purposes
of his brother. Fifty years later he said:

> My brother decided that, by keeping costs of printing and distri-
> bution down to a minimum, and spending "not a dollar for color
> where lines will do,"[3] it would be possible, not only to finance
> his own publications, but also to stimulate research in a domain
> which seemed to be on the threshold of rapid expansion. A more
> canny man would have cautiously felt his way along, soliciting
> editorial cooperation in influential circles, campaigning in advance
> for neurological papers, and securing a guarantee fund adequate
> to meet the inevitable initial deficits. But that was not his way
> of doing things. How much advance correspondence there was
> about the enterprise I do not know; probably very little. Cer-
> tainly there was no editorial collaboration and no financial sup-
> port of any kind.

In all probability it would not have occurred to Clarence
Luther Herrick that he had peers in America to whom he could
turn for advice. There were few senior neurobiologists at that
time, except possibly Burt Wilder at Cornell, who would have been
sympathetic.

But why did Herrick decide to begin publication of the
Journal of Comparative Neurology at that particular time? Possibly
because he found himself being considered for a professorship
on the faculty of the new University of Chicago. Professor
Whitman, also being considered for a post in biology at Chicago,
had founded a journal. Herrick wrote to President Harper on
January 12, 1891, saying that he would expect the university
to provide financial support in establishing a journal as one of
the conditions under which he would consider an offer of a
position. That was at least two months before the first issue of the
Journal went to press and five months before he received the
firm offer of a professorship. Did it help him get it? Perhaps. But
he did not receive a single cent of support for it either from the
University of Chicago or from President Harper, who had
pledged two hundred dollars personally.

If there was anyone in Cincinnati with whom Herrick discussed his plans to start a journal, I suspect it was another one of his students, Charles H. Turner, whose lengthy research paper, "The Morphology of the Avian Brain," was nearing the stage of publication. Turner was the son of the janitor of a Baptist church and later became the first black student to obtain an advanced degree at Cincinnati. As he enjoyed no relationship with Denison University, it may have seemed inappropriate for him to submit his article to the *Bulletin of the Scientific Laboratories of Denison University.*

Having decided to proceed, for whatever reasons, Herrick assumed full responsibility for his *Journal,* entering into a private contract with a local printshop. He assembled manuscripts and illustrated his own with drawings, some of which were reproduced as lithographs from his etchings on stone. He may have had some help from students, but his brother did not recall participating. No articles were solicited for the first issue.

Charles Judson Herrick[4] retrospectively expressed the belief that his brother expected the *Journal* to become a clearinghouse for ideas and research findings in the nervous system: "It was his hope that from fusion of diverse interests and technical skills the puzzling problems of nervous physiology would be resolved and that these researches would finally yield a solution of the biggest puzzle of all—the mechanism of the organic relationship between the physiologically observable functions of the nervous system and those mental processes that can be known only subjectively, i.e., the mind-body problem."

Policy and scope of the *Journal* were stated in an editorial in the first issue:

The problems of neurology resolve themselves into the purely structural investigation, which appeals to microscope and microtome, and physiological questions involving a knowledge of the behavior of the living cell under the most diverse conditions, as well as of the laws of composition of function due to their interaction. Yet a higher class of problems, which properly transcend the sphere of neurology, as of all purely observational science, respecting the relation of body and mind, can never be wholly ignored. . . . That part of the field which is being cultivated with the most zeal and success is the structural province. Yet in this most promising department the accumulation of details has too often proven unfruitful for the lack of a sufficiently comprehensive view of the entire field to enable the investigator to appreciate the bearings of isolated facts.

Early in 1892, when Herrick was in Berlin, he had an announcement printed as a handbill. It contained other statements of policy:

> In addition to anatomical and physiological papers, there will be special attention given to habits, instincts, expression of emotion— in short, all data germain to a true comparative psychology. . . . The present situation of the science is destructive to critically accurate investigation so long as it is necessary to seek with great effort and no guaranty of success through the vast list of zoological, anatomical, physiological, general biological, and other journals and society-reports for the fragmentary literature of neurology.[5] All who appreciate the necessity for some authoritative and complete magazine for this important department are earnestly requested to assist in making it truly international in character. . . . Contributors are requested to observe that articles will be printed in English, German, or French but, unless especially otherwise requested, manuscript in European languages will be done into English by the editors.

It is apparent that the *Journal of Comparative Neurology* was expected to cover the field that we now designate as neuroscience.

The first volume of the *Journal* was printed in Cincinnati and while Clarence Herrick was absent in Europe his brother, Charles Judson, having moved to Granville, encountered difficulties in maintaining liaison with the Cincinnati printer. The title page for Volume I was printed in Berlin and sent to Granville to be added to the assembled numbers. It is reproduced opposite. The *Journal* was mailed to subscribers from the Granville post office. Subsequent volumes were printed by a shop in Granville where the editors could oversee the work.

A notice in the May number of Volume II stated that "protracted absence [of the editor] in Europe has delayed the present number considerably and will also account for a number of unfortunate omissions and errors in the special February fascicle." There were graver problems in store. That the project succeeded at all was due to the determined efforts of Clarence with loyal support of his brother Charles and a considerable amount of money from their own meager academic salaries.

Clarence Luther Herrick was able to maintain active editorship of the *Journal* for only three years, until his health broke in the winter of 1893, at which time his brother, of necessity, assumed major responsibility.

An insight of the activities of the founder can be gained by examining early volumes of the *Journal*. It did not take long for

THE JOURNAL

OF

Comparative Neurology.

A QUARTERLY PERIODICAL

DEVOTED TO THE

Comparative Study of the Nervous System.

EDITED BY

C. L. HERRICK,

PROFESSOR IN THE BIOLOGICAL DEPARTMENT OF CHICAGO UNIVERSITY,
LATE OF THE UNIVERSITY OF CINCINNATI.

Volume I.

Price 3,50 DOLLARS per Annum.

Granville, Ohio, U.S.A.

R. Friedländer and Son, Berlin, European Agents.

contributions to arrive. The first volume, completed in December, 1891, contained seven original articles: "Contributions to the Comparative Morphology of the Central Nervous System Parts I, II, III" and "Contributions to the Morphology of the Brain of Bony Fishes, Part I" by Herrick and his brother; "Morphology of the Avian Brain" by his student C. H. Turner; "Recent Investigations on Structure and Relations of the Optic Thalami" by H. R. Pemberton; "The Morphological Importance of the Membranous or Other Thin Portions of the Parieties of the Encephalic Cavities" by Burt G. Wilder; "The Arachnoid of the Brain" by J. W. Langdon; and "The Lumbar, the Sacral, and the Coccygeal Nerves of the Domestic Cat" by J. B. Stowell. There was a translation by Oliver S. Strong of a paper by C. von Kupffer, "The Development of the Cranial Nerves of Vertebrates"; and short notes entitled "Metamerism of the Vertebrate Head" and "Laboratory Technique" by the editor, one note describing an operating table for use with dogs as subject of experimentation without anesthesia.

Volume I contained also two editorials, "Problems of Comparative Neurology" and "Neurology and Psychology." The former provided an overview of the subject based mainly on recent work of foreign investigators, notes on silver staining by methods of Golgi and Ramón y Cajal, cerebral localization techniques, metamerism, and the cranial nerves. In the second editorial, Herrick reviewed developments of the preceding twenty years (back to 1870) and sketched the movement toward a physiological psychology. He noted that "psychology, after long hesitating to avail itself of the help . . . offered, has apparently been able to do little more than clothe itself in the garb and acquire the language of neurophysiology."

Remarkable features of Volume I were the sections "Literary Notices" and "Recent Literature," mainly current (1889-1891). There were 21 of the former and 284 references. There is little doubt that some of them were found in journals received in exchange for copies of the *Bulletin of the Scientific Laboratories of Denison University,* others in those of Cincinnati libraries.

Volume I concluded with a brief announcement that correspondence should be directed to Granville, Ohio, while the editor was in Europe, and that after October, 1892, the *Journal* would be issued regularly from the University of Chicago. But, as we shall soon see, that never came to pass.

Volume II began while Herrick was in Berlin in 1892, the first number being dated February. His contributions to it were based on research conducted at the University of Cincinnati. One article was entitled "The Cerebrum and Olfactories of the Opossum, *Didelphys Virginica,*" and in it the author drew comparisons with his previous observations on brains of the higher mammals. Another article, "Contributions to the Morphology of the Brain of Bony Fishes, II," appeared in the May number.

The final number of Volume II, December, 1892, contained a study entitled "Embryological Notes on the Brain of the Snake." This sortie into embryology represented an approach to neurological research different from that with which he had been engaged in Cincinnati; the article was completed after Herrick resumed work at Denison University.

By the time Volume III appeared, Herrick had been able to conduct some experiments in Granville. The first number contained the article "The Development of Medullated Nerve Fibers." Other issues contained contributions to the series on comparative morphology of the brain, based mainly on his work in Cincinnati.

After Clarence Herrick's health broke, his brother Charles Judson assumed major responsibility for the *Journal* and was listed as coeditor of Volume IV and subsequent volumes. Clarence had been granted leave from Denison to try to regain health in New Mexico and felt an obligation to continue making contributions to the *Journal,* but only one of them (in Volume VIII) was based on research conducted in the Southwest.

He watched with interest the changes in his *Journal.* In May, 1904, he commented on its reorganization as the *Journal of Comparative Neurology and Psychology* with Robert Yerkes added as a coeditor: "There will be many who will see little sense in the 'behavior' matter and I confess that results so far are dreadfully disappointing. I think the behavior people have got to become histologists and pathologists or do their work in collaboration with such before the results will be of any great use." His student George E. Coghill would attempt to do just that. The name *Journal of Comparative Neurology* was resumed with Volume XXI in 1911.

The Journal of Comparative Neurology succeeded in spite of many discouragements. There were only forty-four paid subscriptions to the first volume and the total income for the three volumes edited by Clarence Luther Herrick amounted to only about $400, but no financial records were kept and on many occasions bills for indebtedness were simply paid out of his pocket without any accounting. Many copies of the *Journal* were exchanged and a stock of them was stored in the attic of Barney Science Hall in Granville. A fire in 1905 destroyed all of the latter as well as Herrick's library of periodicals received by exchange. The founding editor was no longer living when that occurred.

NOTES

1. C. J. Herrick, The founder and the early history of the Journal, *J. Comp. Neurol.* 74: 25-38, 1941. See also C. J. Herrick, One hundred volumes of the *Journal of Comparative Neurology, J. Comp. Neurol.* 100: 717-56, 1954.
2. Volume 74 of the *Journal of Comparative Neurology* contains essays by C. U. Ariens Kappers, G. E. Coghill, and A. Meyer.
3. He may have been thinking of the *Journal of Morphology*.
4. C. J. Herrick, 1955.
5. Three years earlier, Herrick had written an essay, "Science in Utopia" (*Amer. Nat.* 22: 689-702, 1888), in which he reported how scientists in that mythical country had been faced with the multiplication of scientific reports: "Under the old system, which closely resembles our own, there was neither official supervision nor recognized limitations upon publication. The great mass of literature soon made specialization necessary with constantly narrowing limits, until the broader purposes of scientific study were rapidly being lost sight of in the attempt to meet the bibliographic obligations thus imposed." Little could the prophet Herrick know about what was in store!

VIII

CHICAGO

In 1890 a successful move was made by a group of citizens to establish a university in Chicago along with their preparations for the World's Columbian Exposition. Marshall Field donated a tract of land adjacent to the exposition grounds and pledges of money, many of them to match a large contribution by John D. Rockefeller, enabled an organizing committee to proceed. They elected one of their own members, thirty-four-year-old William Rainey Harper, as president of the University of Chicago and charged him with the task of securing a faculty in time to begin classes in the autumn of 1892.

Harper had obtained a Ph.D. degree from Yale in 1875 and was Professor of Semitic Languages in that university in 1890 when he took a similar chair along with the presidency at Chicago. Previously he had served as principal of the Preparatory Department of Denison University, from which post he moved to the Baptist Union Theological Seminary of Chicago. There he held the professorship of Hebrew for seven years before being called to Yale.

With the backing of such influential men as Field and Rockefeller, President Harper set out to build a university that he hoped would be second to none in America. He had been given practically a free hand and, even though all the money that would be required was not immediately available, he was confident that it would soon be in hand and he could proceed to recruit faculty on attractive terms. There were some early doubts that his methods of recruitment would prove successful, critics likening them to the autocratic methods employed in the Standard Oil Company by the principal benefactor, but regardless of the means he employed, the ends were impressive.[1]

Harper's first appointment in science was that of Clarence Luther Herrick. Seeking a man who was at the same time an

73

earnest Christian and a scientist of recognized accomplishments, he had been advised by Baptist acquaintances that Herrick was such a person, one who could be counted on to avoid conflict between religion and the new biology with its concepts of evolution.[2] Harper had experienced some criticism for his own effort toward "scientific" study of the Bible and was aware of the danger of agnosticism arising in his science departments. Although he had no knowledge in the realm of biology, he recognized that it should be strongly represented in his university. Finding a rising young biological scientist of the Baptist faith was indeed a stroke of luck.

President Harper opened negotiations with Herrick by letter on January 6, 1891, asking him whether he might be able to consider a position should one become available in the new University of Chicago. Although in later years Herrick recalled: "I replied that I was well pleased with the outlook where I was and had no taste for the work of organization, being intent on research." His actual reply, dated January 12, 1891, expressed considerably less disinterest. Among his comments in that letter were the following: "I have long felt that scarcely anything is more essential to the evangelization of our time than the union of religion and the highest type of intellectual culture. . . ." He said that he missed "the direct Christian fellowship and activity which is absent here [in Cincinnati]." These declarations must have been good news for President Harper.

Herrick contemplated his own role there to be that of a teacher of undergraduate courses, in which he had had six years of experience. He told the president that he would consider an offer if it provided opportunity for him to develop his research and teaching along lines of physiological psychology, histology, embryology, and comparative neurology. He emphasized the point that he was "primarily a biologist." As a further consideration, he would expect support from the university in launching the publication of a journal devoted to comparative neurology.

Herrick may have entertained doubt that all of his demands would be attractive to Harper, and closed his letter with a warning that should the organization of the new University of Chicago produce "subjective limitations" of the lines he wished to follow, he would withdraw his name.

Months passed without word from Chicago, and Herrick began to wonder if he was still under consideration. He dispatched another letter to President Harper on April 27, 1891, asking whether there had been any new developments. He told Harper that the University of Cincinnati had increased his allowance for equipment and that the first issue of a new *Journal of Comparative Neurology* had been printed.[3] He declared that he could not jeopardize a promising future at Cincinnati for an indefinite prospect at Chicago.

Negotiations with other scientists for positions in the new university were underway at that time. Following the German organizational pattern, each major department was to have a "head professor." The head professorship in biology was soon to be offered to C. O. Whitman of Clark University. At that moment, however, Harper encountered delays in the availability of money from the estate of a Chicago benefactor and was unable to make firm commitments for biological buildings and equipment. He was anxious to keep negotiations with Herrick open and asked for an appointment to meet and discuss matters with him during a forthcoming religious conference in Cincinnati.

They met early in June and, without arriving at any definite answers to questions of title, money for equipment, nor subsidy of the *Journal,* Harper encouraged Herrick by asking him to draw up an outline plan to indicate the nature of the work he wished to pursue, classes he expected to teach, and a statement of his requirements. He also asked for names of other scientists who might be approached for appointments in biology. He had requested similar information from Professor Whitman on June 2, 1891. Herrick complied with Harper's requests and sent him the first of several memoranda, which was based on his expectation that his work would be to a large extent undergraduate teaching.[4] Harper was satisfied and on June 15, 1891, responded with an offer.

Although this offer was vague, Herrick grasped it with some enthusiasm. Being content to rest his future on the word of a

Christian gentleman, he presented his resignation to the University of Cincinnati, effective at the end of 1891. It was regretfully accepted.

There were more exchanges of letters. Herrick decided to go to Europe to prepare for his new professorship and was encouraged by Harper to purchase equipment for his new laboratory. He took his family back to Granville, where they were to remain while he was abroad, and borrowed money for his travel to Berlin.

During his communications with President Harper, Herrick came to realize that his department in the University of Chicago was to be one in which graduate studies would be given priority over the usual college courses. He replanned his program and before sailing for Europe sent the president a revised and expanded outline.[5] This clearly documents Herrick's proposal for inter-disciplinary studies in neuroscience. Because of its relevance to the genesis of American neuroscience, the outline is reproduced in Appendix II on page 131.

This memorandum was accompanied by a list of equipment which reveals that Herrick expected his department to encompass much of neuroscience. He requested, for example, microtomes and microscopes for histology, incubators for embryology, kymographs and electrical stimulators for physiology, craniometers for physical anthropology, equipment for photographic darkroom, machine tools, animal cages and aquaria, and "apparatus for accumulating vital statistics." He was indeed planning an institute of neuroscience.

The academic personnel were to consist of the professor in charge, a docent (neurology), and fellows in physiological and comparative psychology. He recommended appointment of "Mr. C. H. Turner, whose recent papers in the *Journal of Comparative Neurology* have attracted much favorable attention," to be superintendent of the animal house. Herrick proposed for him a salary of $500-600 per year and living quarters in the animal house. He called him "the ablest colored man I have ever seen and a gentleman . . . a Baptist and (what is strange) I think a Christian." I wonder just what Herrick's criteria were for questioning whether this son of a Baptist preacher was a Christian. His attitude toward

Blacks must have been the general feeling of the time in Cincinnati.

President Harper wrote again before Herrick sailed for Europe:

> Your favors of November 21st and 23rd [1891] have been received. I am much obliged to you for all they contain. I understand that you are to work along the lines you indicated. I may say to you that we are negotiating with Professor Whitman of Clark University for the headship of the Department of Biology. The plans are by no means clearly laid out but you will be protected in the particular department of work in which you wish to labor. The laboratory will of course be constructed from a very broad point of view and will contain everything which you suggest. How soon the laboratory can be planned in detail I cannot now say. I hope you will not think me too indefinite in this matter, but believe that at present it does not seem possible to write more definitely. I trust that you will have a most prosperous time abroad and that your work will be all and more than you expect. Perhaps I may hear from you again before you leave the country.

Herrick departed from New York after Christmas, 1891, with this comforting assurance that everything would be provided to effect his dream of an integrated program in neuroscience. He had no objection to the appointment of Whitman, whom he recognized as a distinguished biologist, sixteen years his senior, recipient of a Ph.D. from Leipzig, director of the Woods Hole Laboratory, and founder of the *Journal of Morphology*. What he did not know was that Whitman, in the position of Head Professor and member of the University Senate, would be able to exert powerful influence over President Harper in respect to appointments.

An unprecedented opportunity to acquire illustrious faculty for the University of Chicago had been called to Harper's attention sometime between June and December, 1891 (probably by Whitman). The president of Clark University, psychologist G. Stanley Hall, had encountered serious financial difficulty due to unfulfilled pledges of support. The Clark faculty role was overextended; many of the members, faced with the prospect of drastic reduction in salaries, had resigned and the trustees had accepted their resignations on January 21, 1892. President Hall recalled that "very soon after this, President Harper . . . appeared on the scene." Professor Whitman, who had decided to join the University of Chicago, was only the first of fifteen of the Clark faculty whom Harper was able to recruit.[6]

Another member of the Clark faculty lured away was Charles A. Strong, a brilliant twenty-nine-year-old docent in psychology who had a bachelor of arts degree from Harvard but no advanced degrees. Strong was the son-in-law of John D. Rockefeller—reason enough for Harper's interest in him.

Herrick had been in Berlin only a few weeks when he became suspicious that all at Chicago was not going along lines of his own best interest. Gossip among Americans in Berlin raised doubts. He wrote to his brother Charles on the last day of January, 1892, saying that he had met "Mr. Tufts who is to teach in Psychology [he meant philosophy] department of Chicago Univ. . . . I learn that Strong, son-in-law of Rockefeller, is to have charge of some work in Phys. Psychology. That is where Harper's hitch comes in the matter of my department no doubt." By that time he had received a long-delayed letter from Harper that had been written December 14, 1891,[7] in which the president told him "the title of your chair is still unsettled. . . . If you are willing to let it stand neurology there can be no objection." Herrick did not yet realize that the physiological psychology had been taken away from him, but he had no objection to the change in title of his professorship. He promptly gave the byline for an article in the *Anatomische Anzeiger* as "Professor of Neurology in the University of Chicago."

There were other disturbing statements in the same letter from President Harper. He said:

> I do not quite see our way to guarantee $500.00 for the *Journal* until matters are in shape. [He did pledge $200 from his own pocket.] Would it not be possible to let it fall behind a little without any real injury? . . . It seems altogether certain that we shall be able to authorize you to purchase a small outfit of microscopes, etc., for the University and yet here again we are not in a position to take action because we have not reached that point. If you are willing to advance the money for them I am sure we shall be glad to take them off your hands, but we have no plans as yet in this direction.

Herrick thought about these matters and on February 3, 1892, seemingly in an agreeable mood, wrote to Harper again. The full significance of the appointment of Mr. Strong had not yet dawned on him. He still wanted to maintain faith in the word of the president and he commented: "I recall at this moment that you expressed yourself as more perplexed as to my department than any other and since I have been here one or two things have occurred to me which perhaps it will yet be time enough to consider. . . ." He continued with a play of words: "It may be that it will seem best to have a Strong Psychology Department. . . ." At that point a professorship in neurology was satisfactory to Herrick.

President Harper had begun his negotiations with Strong about the middle of December, 1891, before Herrick left America, without telling Strong that he had promised the laboratory of physiological psychology to Professor Herrick. In addition, he acceded to Strong's wishes to have the anatomy and physiology of the nervous system[8] under his jurisdiction as well. There was clearly a conflict over two of the principal fields the president had promised to protect for his first-chosen professor.

To the other matters in Harper's letter of December, 1891, Herrick responded that he could not afford to purchase much equipment for the University of Chicago with his own borrowed capital. He asked Harper to send him the promised $200 to enable him to get out the March issue of the *Journal of Comparative Neurology.*

An official notification of his appointment reached Herrick about that time. It was a longhand letter from T. W. Goodspeed, Secretary, and is reproduced on page 80. It was not in neurology, but in biology, that he was appointed. And he would be without salary for nine months.

Harper's perfidy became clearer to Herrick when he received a brief letter containing the off-hand comment: "I hope that you and Prof. Strong are in correspondence with each other and that the result will be to your mutual satisfaction." They had been; and it was not! Charles Strong had sent a cordial letter to Herrick on February 29, 1892, at Harper's request. Herrick replied

The University of Chicago.

Office: 1212 CHAMBER OF COMMERCE.

Chicago, _____ Jan . 30th 1892

Prof. C. L. Herrick.

My Dear Sir,

At a meeting of the Board of Trustees of The University of Chicago held yesterday, Jan. 29th you were elected Professor in the department of Biology. The Salary will be $3,000. per year. payable monthly, at the end of each month. It is understood that your term of service will begin Oct. 1. 1892, and that the Salary above named begins at that time.

Hoping to receive an early acceptance of this position on these terms, I remain,

Yours very Truly,

T. W. Goodspeed,

Secretary.

coldly to Strong on March 30, 1892: "While this is the first intelligence of a cancellation of an arrangement made in May, 1891, I was not wholly unprepared for it. . . . It appears, however, that there is a third person, name to me unknown [Loeb?] who supposes he has part or all of physiological psychology in charge."

The more he thought about the matter, the angrier Herrick became. He had dispatched a conditional resignation to the University of Chicago trustees on March 14, 1892:

> Gentlemen:
>   In view of an ambiguity respecting the nature of the work and status of the chair to which I have the honor to be elected I crave permission to withdraw my acceptance of the same either definitely or pending more complete deliniation of the scope of the chair. With highest appreciation of the honor I find myself reluctantly obliged to forego
>
> > I am very respectfully yours,
> > C. L. Herrick

By the same post he sent President Harper a letter in which he said:

> I wish to meet you personally before expressing an opinion as to the propriety of permitting me to come to Europe at my own expense to prepare for work which had long since been allocated to another . . . on any supposition now available I have no independent status. . . . From the personal stand point my situation is disastrous. . . . I will return to America at once and seek to undo the mistakes I have been making. . . . Meanwhile I send you herewith my resignation of the position to which I was elected at your suggestion.

President Harper must surely have realized that he had broken a legal contract with Professor Herrick. Moreover, it may be assumed that he did not relish the thought of his first-chosen professor, a distinguished scientist and an earnest Baptist, departing under circumstances that could place a stigma on the president of the University of Chicago. His aim, then, was to try to placate Herrick and maneuver him into a subordinate position.

Charles Strong, upon learning about the duplication of appointments, sought to relinquish part of his work in psychology, and give up the laboratory to Herrick, but Harper refused to make any change that would affect the title granted Mr. Rockefeller's son-in-law. On the other hand, he declined to accept Herrick's resignation, dispatching a long letter to him in Berlin on March 28, 1892, from which the following comments are extracted:

> I have not understood that in any arrangements made with Mr. Strong, there was a cancelling of arrangements made with you. . . . Will you allow me to say just a word? From the moment I was appointed president of the University of Chicago, I made up my mind that if it was possible to secure *you*, we wanted *you* in the University. . . . I wish to assure you that in anything which has been done we have certainly no desire to interfere with your plans. . . . You are aware I am sure of the peculiar reasons why Mr. Strong's presence in the University is desired.

Herrick replied on April 11, 1892, from Berlin:

> Yours of March 28th is just received reaching me the day before my departure for Hamburg. I regret that my letter caused any trouble. I desire above all things to cause no one else trouble in the matter about which I have had sufficient for all. I congratulate you on the bright plans for the Biological department. As to my own relation to it, probably it will be best to leave it an open question for the present. I have been too much shaken up to consider the possibilities satisfactorily. In fact, have been canvassing the desirability of seeking some other line of work which might bring me less directly into contact with Christian people. Perhaps if I had known you personally & had not known a good many people in Cincinnati quite so well I could have taken more for granted. It is at any rate a great satisfaction to know that all has been done with the kindest intentions. Having been ill and greatly irritated by petty perplexities & financial hobgoblins I am perhaps in no mood to calmly plan for the future. I trust Prof. Strong understands that under no circumstances will I infringe in any way or to any extent on his territory but on the contrary shall take pleasure in being of any assistance possible. It will be quite out of the question in any event for me to do any work in Physiological Psychology. The news that Prof. Whitman has accepted the head of the department will, I think, be received with general satisfaction. Will reach Granville about the 26th D.V.

after which there may be opportunity to confer personally which will perhaps be most satisfactory.

He then added a long postcript:

In again reading your letter I am impressed that you regard my position as incomprehensible not to say unreasonable. I certainly am not mistaken that in our first interview when I remarked that I was working toward Physiological Psychology via nerve anatomy you asked me if I could not be ready for it and said a strong department was desired with outfit & so forth. I subsequently answered in the affirmative & have had no hint from you that a change of plan had been made. Although you did say in a letter before I left America that my place was giving more trouble than any other. My only avenue of information respecting the university & its personnel has been your letters & the press otherwise I should have understood your hint & saved myself a serious blunder. If you had frankly stated the nature of the perplexity there could not have been a moment's trouble for the comparative aspect is ample. I say this much to explain myself. I have suffered a great deal during my school and teaching experience by being on friendly terms with conflicting elements & have had my confidence sorely shaken in human nature by the bitterness due to clashing interests etc. & there is not money enough in America to induce me to put myself before students in such a light as many of my former teachers stood in. My resignation is solely on the ground that I will not even *seem* to enter into competition for a place or interfere in the plans of another. I have entertained great doubts as to my adaptibility for the kind of work incident to the founding of a large institution. I can understand that it is desirable to have some of the very few Baptists who have attempted to work in science associated with Chicago Univ. but otherwise perhaps Prof. Whitman may arrange the force quite well without one. At any rate there can be no harm since a change of footing is necessary to let things shape themselves according to the demands of the case. I hope you will not inconvenience yourself on my account. Something will open—as for the degree—that was the only foolish notion my wife ever had that I know of. There never was any sense in gilding an ass' hoofs.

Herrick returned post haste to Granville, Ohio. President Harper continued to try to placate him and succeeded in getting him to come to Chicago on May 9, 1892, to discuss the affair. Whether any attempt was made to hold a meeting with Professor Whitman is doubtful. The "rape of Clark University" had taken place a month previously and Whitman had all the professors he wanted in his Department of Biology. Probably no genuine reconciliation could have taken place at that late date.

Before deciding to go to Chicago to meet Harper, Herrick had been admonished by friends (and possibly his wife) for acting "with undue precipitation and irritation." He became willing to reconsider the matter and wrote to Harper on May 2, 1892: "I am quite ready to adjust myself to the changes in plan in any reasonable way. . . . If my resignation has been accepted please inform me at once. . . . If, on the other hand, an independent chair of Neurology is contemplated . . . I should be glad to know. . . ." But it was too late. That post too had been given away by Harper.

While Herrick waited in an anteroom for his conference with President Harper, he saw pictures of newly appointed faculty members on the walls. Among them was one of a man he had considered a friend, Henry H. Donaldson, labeled "Professor of Neurology." President Hall of Clark University had released Donaldson to the University of Chicago and Professor Whitman demanded his appointment to the Chair of Neurology. Harper tried weakly to defend this last post that had been promised to Herrick but, upon Whitman's insistence, he gave in and agreed to relinquish it if he could get Herrick to consent to accept another position.

Years later Herrick recalled that when he faced Harper on the Donaldson issue he was told: "We had to have Whitman and he insisted that the whole corps of Clark should come or none."

Herrick, surprisingly, did not react angrily, but meekly listened to the president's proposal to create for him a new department of comparative psychology. A memorandum, dated May 9, 1892, was drawn, omitting any references to neurology or support for the *Journal of Comparative Neurology*. It was signed by Harper alone and Herrick's copy had that signature removed.

Matters had reached the point of no return. The Donaldson

appointment[9] was the final blow, and Harper only rubbed salt into the wound by suggesting that Herrick develop a section in the Department of Philosophy.

Later, on May 20, 1892, Harper, still trying to make some amends, offered Herrick physiological psychology, which he said Strong had agreed to relinquish, and proposed to combine it with comparative psychology.[10] But neurology was out of reach. He begged Herrick to come again to Chicago, this time at the university's expense. Herrick replied on May 23 that he failed to see that the proposal altered "the situation in those particulars about which I should be most solicitous. . . . I beg you to consider me out of the field and shall esteem it a favor if my resignation may be promptly accepted. . . . It would be embarrassing to me, especially as the earliest appointee, to occupy an isolated and with reference to the Biol. Dept. subordinate position." He closed this letter by wishing President Harper "success in all honorable undertakings. . . ."

Harper replied on June 2, 1892, that he was "very much disappointed, but will lay your letter before the Committee at its next meeting and notify you of the results." However, the door was still open a crack, for Harper went on to say, ". . . I venture to express the hope that before the final settlement you will consent to come to Chicago at the expense of the Committee. Their meeting will be held next Monday afternoon."

Years later Herrick recalled that Harper promised "all sorts of things, among others two years on leave for research with full pay. He said to me that as all the other members of the biological faculty were agnostics it was especially desirable that there should be one 'orthodox' member."

Herrick wrote to Harper once more on June 3:

> My reasons for being unwilling to accept work in the University of Chicago under conditions so different from those contemplated in our original agreement have been stated perhaps as fully as is necessary; I may add however that I am unable to feel that I could count with any assurance upon considerate or honorable dealing from the authorities of that institution. . . . It is not clear to me that I could be in sympathy with the methods & tendencies of the institution as I now understand them.

Harper finally recognized that the end had come. Insofar as possible he proceeded to put the matter out of mind. The Board of Trustees had accepted Herrick's resignation on June 7 and Harper sent him a letter of regret on June 8, which crossed in the mail a final letter of Herrick to the secretary of the Board of Trustees.[11]

Granville, O.
June 9th, 1892

T. R. Goodspeed D.D.

Dear Sir:

I feel that the serious loss and humiliation I have sustained at the hands of the authorities of the University of Chicago justify me in again insisting that prompt action may be taken on my resignation which has for several weeks been in your hands.

Yours very truly,
C. L. Herrick

No reimbursement was made for Herrick's expenditures in Berlin. So ended tragically the dream of a man who wanted to establish neuroscience in Chicago, a man who was far ahead of his time. Not until 1955 was the story told, and then incompletely, by Charles Judson Herrick.[12]

NOTES

1. Harper's 1903 report in *The Decennial Publications, First Series,* Vol. 1, University of Chicago Press, contained the following statement on page xviii: "Two policies were open for the organization of the staff of instruction. The first, strongly urged by many educators, was that of selecting a few younger instructors and allowing the work to grow more gradually under the domination of a single spirit. The other policy, which was regarded as impracticable by many, was one adopted, namely, to bring together the largest possible number of men who had already shown their strength in their several departments, each of whom, representing a different

training and a different set of ideas, would contribute much to the ultimate constitution of the University. Considerable risk attended the adopting the second policy, for it was an open question whether with so large a number of eminent men, each maintaining his own ideas, there could be secured even in a long time that unity of spirit without which an institution could not prosper. During the first year there were times when to some it seemed doubtful if the experiment of bringing together a large number of strong men would prove successful; but during the middle of the second year certain events occurred which led up to the birth, as it were, of the spirit of unity which had not been hoped for."

2. A letter from J. L. Cheney of Ypsilanti, Michigan, dated September 3, 1890, contained the following: "In Cincinnati, as in Minneapolis, his faithful devotion to our Baptist interests has made him no small place in our people's hearts. A first class scientist, who is an earnest Christian, is a rarity. . . . I have thought how great a treasure he w'd be in your new undertaking."

3. Here is a possible explanation of Herrick's sudden plunge into the publication of the *Journal*. He may have believed his case with Harper would be strengthened by doing so.

4. Material in an undated letter of Herrick to Harper in June, 1891, was summarized by Blake (1966): "Herrick's proposed Biology Section consisted of three departments: anatomy and physiology, zoology, and botany. Under one man responsible for the administration and general oversight of the Section, each department would direct the instruction of its undergraduate and graduate students. Characterizing his system as one of 'connected and progressive development,' Herrick proposed that students be introduced to the elementary principles of the science in the college and to the advanced concepts in the graduate school. In the Department of Anatomy and Physiology, which he proposed to take hold of, he would offer elementary courses in anatomy and physiology, and advanced courses in neurology, comparative psychology and physiological psychology. In neurol-

ogy he would deal with the structure of the nervous system; in comparative psychology with 'the expression of emotion or intelligence as exhibited by all animals'; in physiological psychology with the connection between physical and psychic phenomena, or as he phrased it, the relationship between the body and the soul. In the Departments of Zoology and of Botany, he proposed elementary work in the college and an experimental station for graduate study. Each instructor was to have total responsibility for a particular field, from its elementary stages to its most advanced level. The benefit that he saw in such an arrangement was that it gave each instructor access to all the biological students, permitting him to observe and cultivate any who appeared to be especially promising. The advantage to the student would be that he might more easily discover the unity of a particular field and find his interest for independent study quickened."

5. The copy at Lawrence, Kansas, is handwritten. I believe it to be the rough draft of the memorandum to Harper.

6. Harper visited Professor C. O. Whitman at his home and there met with nine of the dissident faculty members, seven of whom later went to Chicago; in addition to Whitman, they were George Bauer, Oskar Bolga, Henry H. Donaldson, Franklin P. Mall, Albert A. Michelson and John U. Nef. He visited President G. Stanley Hall the next morning and offered him a position. Hall later (1923) described their discussion: "I replied that it was an act of wreckage for us comparable to anything that the worst trust had ever attempted against its competitors but he asked, 'What could I do?' recounting the . . . difficulties he had had in gathering a staff. I finally told him that if he would revise his list, releasing a few of our men and taking one or two others whom he had omitted, I would bear the calamity silently and with what grace I could. . . . To this he demured, and I finally threatened, unless he would make such revisions of his list as I suggested, to make a formal appeal to the public and to Mr. Rockefeller himself to see if this trust magnate . . . would justify such an

assassination of an institution as had that day been attempted here (for Harper had made advances to nearly all our staff, even those who remained loyal, and was evidently ready to make a clean sweep). He finally assented, even taking at least one man here who covered the exact field of another he had previously engaged. . . . G. S. Hall, *Life and Confessions of a Psychologist* (New York: Appleton, 1923).

7. It is strange that Harper did not send this letter to Ohio. Did he not wish to raise doubt while Herrick was still in the United States?

8. There may have been another conflict of interest, in neurophysiology, which Harper had promised also to Jacques Loeb, whom he had recruited from Bryn Mawr College for the chair of Physiology.

9. Herrick continued to believe that Donaldson, whom he had considered a friend, was involved, along with Whitman and Harper, in the perfidious treatment.

10. Harper's letter of May 20, 1892, to Herrick contained the following: "Without any communication whatever between Prof. Strong and myself, he sent me last week a list of courses in the Department of Philosophy for next year. I had already sent him a list of those offered by Prof. Tufts. In the arrangement of his own work he has absolutely omitted anything upon the subject of physiological psychology, thus showing his idea which was mine also, that you were to take hold of this work." This was untrue. Strong did not then relinquish physiological psychology.

11. The first faculty meeting of the University of Chicago was held on September 28, 1892. President Harper opened it with a prayer. [*The University Record, N.S.* 16 (4): 235-37 (October), 1930.]

12. C. J. Herrick (1955).

IX

BERLIN

Herrick's brief residence in Berlin, from January into April, 1892, was not a total loss, although it proved costly and ended in the fiasco of his dealings with the president of the University of Chicago. He sailed on a transatlantic steamship in a joyful mood, believing that he was soon to realize his dreams for neuroscience and confident that he could rely on the word of William Rainey Harper.

Winter was not the best season in Germany, but academic work was at its peak. With his command of the language, he promptly fell into the swing of things. Within a month of his arrival in Berlin he reported to President Harper that

a succession of happy circumstances which I should be tempted to call accidents did not the cause lie in part in scientific connections previously made and in several cases forgotten have placed me in possession of the precise facilities I want & I trust to be able to secure commensurate results. Of course the routine lecture work here is not better than good college work in America, I should say absolutely inferior in many cases, & I am surprised to find many of my friends like Prof Chandler delving away in the vain search for advanced lecture work. But the laboratories, privatim courses, etc., give one access to material of the highest value. Profs Fritsch & Hertwig have given me special facilities and opportunities in the most liberal manner. It is pleasant to find the lapse of years since I was here before not entirely unfruitful of recognition, etc. It is a great surprise to find Physiological Psychology in so rudimentary a condition. But that I shall be able to compensate for with Wundt at Leipzig. Our Junior class in Cincinnati evinced more familiarity with the subject after a brief course than the German students in Prof Ebbinghaus' Uebungen.

To his brother Charles, he wrote at about the same time:

90

"Berlin is bristling with notable names and finely equipped institutions but the real work is done in obscure corners." He was auditing lectures by several professors, commenting on them as follows:

> I have been surprised to see how simple and *in extensio* these people make their lectures. Ebbinghaus devoted an entire hour to illustrating the fact that accommodation of the eye is associated with concepts of bodily motions to produce the idea of distance. Munk took two tedious hours to explain the simplest nerve muscle experiment without a piece of apparatus—but he always stuttered so that he repeated every sentence to enable one to get it. Du Bois-Reymond—the Nestor of Physiological Experimentors—carries great style. A preparator sits on one side & his assistant on the other though the latter has nothing to do. I suppose they expect him to drop dead some day in the lecture room. He limps as though paralytic and covers his mouth with one hand turns his back to his audience and mumbles to an audience of 75 or 125. The experiments are well carried out though the preparator seems to do all the work. The pigeon experiment was artistically even theatrically effective.

One of the main purposes of Herrick's trip to Berlin was to prepare for the courses he expected to offer in neurology and physiological and comparative psychology in the University of Chicago when it should open its doors in the autumn of 1892. There is no evidence that he sought to qualify for an advanced degree, though his wife hoped he would do so and a few months later President Harper would offer to send him back to Germany, which would have made that possible. Herrick did not intend to spend all his time in Berlin but expected to visit Leipzig again in the spring and to go to Italy. He gave up this plan when he found it essential to return to America for a confrontation with Harper.

Herrick was warmly welcomed by leading biological scientists and philosophers at their lectures and laboratories, although not always at their social gatherings. To his brother, he related his experience at one of the latter:

> I enjoyed (or tried to under exceedingly embarrassing circum-

stances) an evening at Prof. Ebbinghaus' an evening or two ago. He is the best man in Psychology probably in Berlin. I called on him & he invited me in to meet an American [Tufts]. I did not understand that it was a gathering but found a room full of company among others Prof. Paulsen another philosophical luminary. . . . I declined wine and raised a regular howl "teetotaler," "temperance" & the like & Ebbinghaus almost literally turned his back on me from that time on, while Tufts took his wine & beer & was well received. . . . I do not feel that I shall have one set of habits for America & another for Germany even for the sake of the fellowship of German savants. I shall probably experiment a little with beer & wine but I will do it privately as a scientific experiment.

Clarence Herrick had spent a student year in Leipzig and must have become aware of German drinking customs. The use of alcoholic beverages did indeed trouble him at that earlier time; in his news article in the *Saint Paul Pioneer Press* of 1882 he told about the professor who addressed his class with a bottle on his lectern. It is apparent that his temperance had prevented him from experiencing the real social life of German university students.

He was not snubbed by all Berlin scientists. Professor Gustav Fritsch was especially cordial. Herrick had been interested in his early work on cerebral localization and wrote a four-page essay on this German neurologist for the second volume of the *Journal of Comparative Neurology,* with which he included Fritsch's portrait. The opening paragraph reads: "Nothing is better calculated to impress the American Student with the wonderful progress which the science of neurology has made during the last twenty years than a visit to European laboratories where he will find the patriarchs of the science still in the thick of the conflict."

In Berlin he kept busy, not only attending lectures, but also seeking to recruit a docent and shopping for equipment for his department at Chicago, and of course promoting his *Journal of Comparative Neurology.* He asked his brother to "send 20 copies of each number of vol. I by express to me here. I shall be able to exchange some and use others in advertising." Moreover, he completed work on several articles. One was a chapter in a

Festschrift zum siebenzigsten Geburtstage Rudolf Leukarts, the Leipzig zoologist under whom he had studied ten years previously.

By March, when he had become suspicious that President Harper was not living up to his promises, a letter to his brother in Granville assumed a tone of intense irritation. Charles had been left in charge of the *Journal* and had encountered difficulties in his dealing with the printshop in Cincinnati. Clarence had received proof sheets that were unsatisfactory and castigated the printer as that "pig-headed fellow." He further declared: "One can not live in the same world with people such as make up the bulk of Cincinnati. I would enjoy going a gunning and shooting indiscriminately everything in human form in that town." He complained to Charles about the inferior quality of lectures in the Berlin institutions, about a change for the worse in Germany since his student days, and he was annoyed that the Royal Library of Berlin had no science periodicals available and was closed in the evening. He was in a bad mood.

Soon after his arrival, Herrick had a title page for the first volume of the *Journal of Comparative Neurology* printed at a shop in Berlin and shipped to Granville, Ohio, where his brother was preparing to mail the last number. This page shows him as "Professor in the Biological Department of Chicago University." He established a Berlin agency for the *Journal* and had a circular printed for distribution as an advertisement, indicating in it the scope and policies of publication. It occurred to him that an advertisement for the new University of Chicago might be appropriate and he suggested this to Harper, but the latter requested him not to have it done.

Herrick met Adolf Meyer shortly before terminating his Berlin residence. The latter was introduced to him in front of the aquarium on Unter den Linden by a mutual friend. Their conversation quickly turned to Meyer's intention to emigrate to the New World where he expected to have to practice medicine for a living while searching for opportunities to continue his research on the nervous system. They discussed a mutual interest in the reptilian brain, and Herrick said he hoped that they would keep in touch with each other. Had it not been for the uncertainty of his own future at that time, he might have offered Meyer the post of Docent in Neurology, which position he had been trying to fill earlier.

They did correspond and after Meyer arrived in America Herrick sent out letters of recommendation of him from Granville. He counseled in a letter to Meyer: "I would not advise you to associate yourself with Chicago University as long as Harper is president. There is likely to be a speedy explosion [wishful thinking?]. The paternalism of the institution is enough to prevent a self-respecting man from working in it."

When Adolf Meyer's dissertation on the reptile brain was published in the *Zeitschrift für wissenschaftliche Zoologie,* Herrick saw that his own work in this field had not been cited. (Herrick's papers were at that time inaccessible to Meyer.) He wrote to him: "I am disappointed, I must frankly say in your paper. You did not think it worthwhile to look upon my papers covering much of the same ground." In spite of this unwarranted criticism, which grieved Meyer, he went on to give him encouragement and they remained friends. Meyer later became a valued colleague and contributor to the *Journal.*

Herrick sent a final letter from Berlin to his brother on March 30, 1892, when a wave of homesickness was overwhelming him. In it he said,

> I am of a very unhappy disposition and have had a very toilsome disappointing and perplexing life so far with prospects of more of the same to the end but in spite of it all I am under the surface a happy man and owe it almost entirely to the home which a true wife has made so. Wherever it has been with a madhouse all about, distraction interwoven pandemonium underneath and the devil to pay (and nothing to pay with) I have always had a home. . . . Were it not for music I should not believe in a life beyond. . . .

He returned to Granville on Monday, April 25, 1892, stopping on the way in Princeton to attend a meeting of morphologists. Professor Whitman was there and they sat together. Herrick recalled in a letter to Whitman several years later: ". . . hearing you say that you had not learned what led me to retire from the

Faculty of Chicago [,] you will forgive me if I thought under the circumstances that it was a polite lie intended to deceive no one."

Reunion with his wife and son must have been a happy, though tearful, occasion; no direct correspondence between them has been preserved. Alice Keith Herrick undoubtedly had much influence on him and encouraged him to reconsider his first rash resignation from the professorship in the University of Chicago which he had fired off to President Harper from Berlin.

X

GRANVILLE AGAIN

Herrick received a warm welcome in Granville upon returning from Berlin in the spring of 1892. Not only did he have staunch friends on the Denison faculty, but some members of the Board of Trustees remembered his notable contributions of earlier years as well as his continued loyalty to the college during the four years of his absence. He addressed the Denison Scientific Association on April 30, reading his paper, "Methods of Biological Study in Germany."

Denison had changed very little, but a new president had been installed while Herrick was in Cincinnati. Daniel B. Purinton was the first president of Denison University who was not an ordained Baptist minister. He was thirty-two years old, a graduate of West Virginia University with a master's degree from that institution in 1876. When he came to Granville, he quietly set about reshaping Denison into a true university composed of several colleges and schools, each with a dean or a principal, but all united by one board of trustees and president. Where his predecessor, President Anderson, had failed, he soon began to achieve success.

President Purinton encountered a number of disturbing situations when he assumed office in 1890, not the least of which was the appalling deficiency in the library. The building was an attractive pseudo-Gothic structure, but its contents lacked the most fundamental reference works. The librarian reported that his funds amounted to $25.13, out of which he had bought a dictionary which, with postage (one could not be purchased locally), left a balance of $13.88! Periodicals obtained in exchange for the *Bulletin of the Scientific Laboratories* were kept in the office of Professor Tight, apparently not entrusted to the librarian.

The new president was quick to appreciate the importance of supporting and strengthening scientific programs. Although his own

96

interests were metaphysical and he believed in the biblical story of creation, he recognized the increasing demand for biological and psychological courses, and gave sincere backing to his science faculty. One of his first acts was to recommend commitment of $300 to Professor Tight for the *Bulletin of the Scientific Laboratories* to insure continued publication.

President Purinton lost no opportunity to stress the need of a science building when he was called upon to speak to Denison alumni throughout the state, using plans that Herrick had obtained a few years earlier. He succeeded. Eugene J. Barney of Dayton, a member of the Board of Trustees, contributed $50,000 for that purpose—nearly twice the amount that would have been required to erect the structure proposed by Herrick in 1888. At the same time, William H. Doane, also a trustee from Dayton, contributed $25,000 for a building to house the preparatory department. These two gifts would relieve the critical shortage of classroom and laboratory space that had developed while enrollment at Denison was passing the five hundred mark. They would make it possible to implement Purinton's plans for a true university.

The Barney and Doane gifts came at the time Herrick returned to Granville, nearly bankrupt and severely depressed mentally, his hopes and plans for a neuroscience program having been dashed by the perfidious acts of Harper at Chicago, against whom he was deeply embittered. Denison's president promptly came to Herrick's rescue. At the June meeting of his trustees he spoke about the need to have more science teachers and informed them: "Fortunately we may find such a one in the person of Prof. C. L. Herrick [whose] services as Prof. of Biology . . . I have every reason to hope can be secured by the Board. I earnestly recommend that immediate action be taken in this case." He obtained approval from the trustees to create a new chair for Herrick. The Department of Geology and Natural History was subdivided: William Tight, Herrick's protégé, was pleased to take an assistant professorship in Geology and Botany, and Herrick was invited to become professor of Biology. The offer was accompanied by a pledge of subsidy for the *Journal of Comparative Neurology*. He accepted in June, his spirits somewhat revived with new hope of salvaging something from the wreckage of his glorious scheme.

The president's plans for a university appealed to Herrick, and the two men worked together to put them into effect. They wasted no time in having work started on the new hall of science. Requirements of Herrick and other members of the science faculty were imparted to the architect. Laboratories for physiological and histological experimentation and for comparative psychology were incorporated into the plans, and there was to be an assembly hall for the Denison Scientific Association. Ground was broken and a cornerstone was laid—with appropriate ceremony—one rainy spring day in 1893. The space requirements of a University graduate program in science would soon be met.

The time appeared to be right to proceed toward implementing such a program at Denison University. The master's degree was already being offered in science as well as the arts, but students were turning elsewhere for more advanced graduate work. More and more American institutions were offering programs leading to the Ph.D. degree.

Purinton and Herrick, with space and facilities improving, saw no reason to delay offering such programs at Denison. To that end, funds for fellowships were found and two of Herrick's Cincinnati students, his brother Charles and Edwin G. Stanley, were brought in as teaching fellows in biology in 1893. Stipends were provided for their services as teachers in the preparatory department, Doane Academy. There were seven graduate students registered that year: five in biology with Herrick, one in chemistry and physics under Cole, and one in McKibben's department of modern language.

Enthusiasm for the new graduate programs spawned another periodical at Denison, mainly for the nonscientific departments. The *Denison Quarterly* was edited by Professor W. H. Johnson and three other members of the faculty "with approval of the faculty and trustees," the latter no doubt providing financial support. Herrick contributed to the first volume an essay in four parts: "The Scope and Methods of Comparative Psychology."

The first issue of the *Quarterly* carried an announcement

of two three-year courses of study leading to the Ph.D. degree, one in philosophy and the other in biology. It is noteworthy that Herrick's courses in comparative neurology and psychology were components of both sequences of study. The first year of the biology sequence was expected to serve requirements for advanced standing in medical colleges.[1]

Charles Judson Herrick had married and in the autumn of 1892 had gone to Ottawa University in Kansas to teach. His brother wrote to him the next spring to ask him to return to Granville, saying that

> there is a strong desire for you to be here at the opening of the new era. Tight is as hearty in it as Dr. Purinton & Prof. Cole. There is a similar feeling among the students. I need not add that I am deeply in earnest in the matter. This year of all years . . . Dr. P. is very eager and says he doesn't care how if only you come. He thinks $800 as much as they would stand but that you could have considerable time. Let me hear from you at once.

He kept urging his brother to come to Granville and join him. He wrote again on April 26, 1893, to advise him about his future:

> It is understoood that applicants for the Ph.D. may do not more than one year elsewhere but that this shall not be the last. I think the second year at Harvard might be better than the first. The financial help will be an important item for of course you realize that there is nothing more to come from Grandmother . . . for some time as things now are. I hope you may be guided to do the best thing. The importance of making this first year of graduate school here boom is apparent if the future is to develop as we hope. The new building will be a beauty. If everything goes we may be able to use Will [their brother]. He could manufacture apparatus.

Clarence Herrick succeeded in his effort to move his brother Charles back into his orbit, eventually to become his successor.

More trouble was brewing in Chicago in the spring of 1893, and Clarence mentioned to his brother that "the Chicago alumni of Denison are protesting against the graduate courses and the faculty in Biology are as I expected, trying to discredit me in every way possible—but we shall not be disheartened thereby."

The spokesman for the Chicago alumni was Frances W. Shepardson who had just become professor of American History at the University of Chicago. He had graduated from Denison University in 1882 and had been editor of the *Granville Times* before going to Chicago. He expressed views of the alumni in letters to Herrick. These letters no longer exist, but Herrick's replies have been preserved and they reveal not only his and presumably Purinton's views about the advanced graduate programs, but also his own personal bitterness toward Harper and the Biology Department at Chicago. Part of his reply to Shepardson on April 7, 1893, follows:

> I learn with surprise that you are antagonizing the effort to extend our courses of study. This not only seems to me a work of super-erogation but inconsistent with your position as alumnus and friend of Denison.
>
> Although we recognize the advantage of money in many directions we do not believe that it will afford the same advantages in some others which nature & its very isolation gives Denison here. We simply propose to use these & gain the impetus of advanced work for its reflex influence on undergraduate work. Our work is in no sense antagonistic to Chicago. Yet we recognize no right which the Univ. of Chicago may claim to exercise over us. After securing the utmost which could be extorted from the friends of Christian education, the Univ. of Chicago has violated every trust & confidence & established a bulwark of agnosticism & immorality which will do vast harm. It is dominated by a man as corrupt and unscrupulous as it is possible to conceive. A man who might with less injustice have attempted my life than to destroy my career & character. You cannot expect me to advocate Chicago as a place for a guileless young man.
>
> I do not criticize you who are there but I suggest that you will do well not to attempt to interfere with the legitimate and modest efforts of your Alma Mater to provide some scientific instruction for those who for any local or other reason do not wish to leave at the close of the senior year.

Shepardson must have written to Herrick again when he

received his letter, and a reply of April 16 from Herrick contained the following:

> I am quite persuaded of your allegiance to the Alma Mater but I still regret the attitude the Chicago alumni are pleased to take in the matter & I think it rests on misapprehension. It is not expected that Denison will open a full Ph.D. course in all departments. Those who may desire to pursue such a course in most lines will probably go *east* or *abroad* wherever the best facilities are offered but it is felt that Denison has been greatly injuring her good name by the promiscuous distribution of honorary (?) degrees. I wonder that the alumni of Denison have not attacked this grotesque evil! What we propose is to offer certain lines of graduate study to those who for whatever reason desire to continue their work here. As to facilities, by the time this goes into effect there will be fair opportunities for the sorts of work offered. I myself purchased by the authorization of Harper a partial outfit for my supposed department in Chicago Univ. This of course came back on my hands by his dishonesty. We shall have much more than Dr. H. proposed to supply me for similar work in Chicago Univ. If you will observe the work offered you will see that the *extras* now proposed are largely in my line. Now as I am engaged for but half the year I am wholly free for half the time for this work. Professor Tight is thereby relieved of much of his work and a large share of my time is reserved for such work as I chose to give it. It all reduces to the question of competency & as the last proposition made me at Chicago was the full Chair of Phys. Psychology—a year in Europe on full pay, & equipment, it looks a little out of place from the latitude from which it eminates without regard to its truth or falsity. . . . Now will you not let this thing work itself out without opposition from the alumni & help us all *you* can?

If anything more was heard from the Chicago alumni, no records survive. Herrick, with full backing of an enthusiastic president and cooperative colleagues on the faculty, forged ahead with his work. He published twenty-seven articles in 1892 and 1893. Most of the studies had been conducted while he was at Cincinnati, or even earlier, and only about half of them were morphological or physiological. He had not had time to get experiments underway in Granville, but on May 4, 1893, he wrote to his brother, saying, "I am putting up a little building with space for a shop on my own lot in which bird skinning and dissecting may go on." This, he expected, would provide the space needed until the new science building should be completed.

Herrick turned to writing philosophical and speculative articles on such topics as the psychological basis of feelings, instinctive traits, scientific utility of dreams, and the soul of man, during the period following his return from Berlin.

The first volume of the *Denison Quarterly* carried an essay by him entitled "Scope and Methods of Comparative Psychology," written with the view of presenting the subject to nonscientific readers, including trustees, some of whom still looked with jaundiced eyes upon the encroachment by biological science into the curriculum of their sectarian college. If studies in physiological and comparative psychology were to be accepted it was essential that they be explained rationally, and Herrick, who was recognized as an earnest Christian, was the person to do that.

Herrick had supported the philosophy of monism as opposed to dualism since the early 1880s. He had written an essay, "Lotze's Ontology—the Problem of Being," for the *Bulletin* in 1889.[2] That was a time when the soul was considered to be something quite real and separate from the body; and Herrick's view of the mind in relation to the human brain seemed by some to be antagonistic. That he incurred criticism is clear from a footnote in one of his later articles in the *Quarterly:* "The writer well remembers how, early in his experience as a college teacher, a member of the board of trustees whom he had unwittingly offended industriously circulated the charge that, in teaching monism, he was propagating materialism."

The trustee may have been Adolphus Behrens, D.D., whose essay in the *Quarterly*[3] concluded:

> Monism is the fashion. Scientists and philosophers are loud in its praises. Theologians are reconstructing their dogmatics along its lines. But for ourselves, so long as it has no support in sober science, and is squarely challenged by a sound psychology, and

involves so many and serious revolutionary influences, we are not prepared even provisionally to adopt it. It is only a new word for an old heresy, than which none has ever wrought more disastrously against morals and religion.

Herrick replied in the next volume of the *Quarterly*[4] with his essay "The Critics of Ethical Monism," in which he said:

Biological speculation and research has been most potent in undermining the last refuges of materialism. . . . It [dualistic orthodoxy] is a theology of demons and the devil who thwart the power and plans of God and who compete for the allegiance of human souls by the use of occult forces. Demonology, Schlatter, Faith cure *et id genus omne* are legitimate products of dualism.

The editor of the *Quarterly* appended a long footnote (seemingly apologetic) in which he stated: "When certain forms of it [monism] come with the recommendation of such names as those of Dr. Strong and Professor Herrick, one cannot afford to turn it away without a hearing."

The significance of this controversy lies in the role Herrick conceived for the mind as an expression of cerebral functions. Physiological psychologists could not exclude the mind or soul as an independent and ethereal thing from their researches on the brain.

Herrick's dream of a multidisciplinary graduate-level program in neuroscience was not fulfilled in his lifetime. Two unfortunate events terminated the dream. The first, and less important one, was a national financial depression which began to put a strain on the university's plan of expansion in 1893. The second was more immediate and disastrous. Herrick came down with a respiratory infection in December, 1893, which culminated in an incapacitating and frightening pulmonary hemorrhage. There being no known treatment for tuberculosis except rest and a dry climate, he was advised to go to New Mexico, but he remained in Granville until summer.

Denison University granted him leave. President Purinton reported to the Board of Trustees, meeting in June, 1894: "All friends of the University regret exceedingly the serious illness of Professor Herrick, who is now absent in the hope of recovering so as to take up his work again in September."

His colleagues and students fully expected him to recover his health and return to the new laboratories that would be awaiting him. Professor Tight (1905) recalled one of the last events before Herrick left:

> He was instrumental in the construction of the new science build-ing, "Barney Memorial Science Hall," yet he never was to work in its laboratories. When overtaken by sickness and it was known that he must leave Denison, some of the "boys" went to his home with a closed carriage and took him to Barney Science Hall, and carried him through the fine laboratories he had so carefully planned, in their arms, and he remarked that he believed that he knew how Moses felt when he was permitted to view the prom-ised land.

Clarence and Alice Herrick later went to Minneapolis, and by July, 1894, Clarence was well enough to travel to New Mexico.

NOTES

1. Soapstone dissecting table tops were stored in the basement of one of the college buildings as late as 1918.
2. C. L. Herrick, Lotze's ontology—the problem of being, *Bull. Sci. Lab. Denison Univ.* 4: 135-38, 1889.
3. A. J. F. Beherns, Monism, *Denison Quart.* 3: 73-79, 1895.
4. C. L. Herrick, The critics of ethical monism, *Denison Quart.* 4: 246-55, 1896.

XI

NEW MEXICO

The pulmonary hemorrhage that struck in December of 1893 was followed by several months of complete incapacitation and depression while Herrick waited out the winter in Granville. His wife gave birth to their second daughter, Mabel, on January 25. By summer he was well enough to travel. Leaving the baby and her little sister, Laura, with their grandmother, Clarence and Alice Herrick took their nine-year-old son, Harry, with them to Minneapolis. The purpose of their trip appears to have been to settle some business matters and to consult with Clarence's friend Thomas Roberts, M.D., regarding therapy and prospects of his recovery. He was encouraged by accounts of improvements of other patients in the arid climate of New Mexico.

Herrick's spirits rose and fell with the state of his health. He wrote to his brother Charles Judson Herrick, who was editing the *Journal of Comparative Neurology* and sharing with William Tight the burden of Clarence's classroom and laboratory work. His letters during July, 1894, reveal the state of his tortured mind. In one dated July 9 he said he could face death "calmly and with some satisfaction, but the thought of sacrificing my dearest ones upon an altar of slow torture has pretty nearly sapped what faith I had left, and I am all but ready to curse God and die." Five days later he wrote:

I was gaining strength and 140 lbs., feeling well and attributing it to guicol [?] when the guizzard [?] violently struck. Heart went on a bump and I did a little unnecessary bleeding. I was at the time living on six to eight raw eggs a day with granum and other imperial drops and felt monarch of all I surveyed. I went back to shaddow soup and toothpicks and have been slowly picking

105

up ever since. The hemorrhage was very small and was only a reminder of what is to be expected. A good deal of mutual worry and excitement over finances, etc., is no doubt the *versa causa* but the papers are all filled and I suppose I am out of it. . . . If my acute disease should assume a chronic type in New Mexico I may spend years there in some other pioneer work but sooner or later each shall go to his own place.

Herrick's last letter from Minneapolis to his brother in Granville was written on July 24. In it he said: "We hope to get off tomorrow. I will have the deed recorded and sent you." He arrived in Albuquerque, New Mexico Territory, on July 29, 1894. His letters throughout the rest of the summer showed optimism over the improvement in his health and expressed some belief that he might be able to return to Granville but should have to spend winters in New Mexico in all probability.

Denison University, in granting him leave of absence with a stipend of $1,600 one year and $1,700 the next, expected that he would be able to return, for his presence was essential to the success of the new graduate program. He continued to hope, and after one year in New Mexico he wrote to his brother to say that he would see him in Granville in October.

President Purinton, too, was optimistic and reported to the trustees in June, 1895:

> I greatly regret the continued absence of Prof. Herrick on account of ill health. His physician has recently decided that he may return to his work with the opening of the school year in September. . . . I respectfully recommend that he be continued as last year at the same salary. Professor Herrick has, during his enforced absence done, a great amount of work in collecting and mounting valuable specimens for the University Museum. In this way we have added to our collection, some 500 birds, 600 minerals, and 2500 botanical specimens.

Herrick did not return to Granville. He wrote again to his brother on September 1, 1895:

> At the last minute I am forced to give up hope of return and with it everything else. It is too bad to leave you in the lurch in

this way but I had fully purposed to go—had trunk ready and some work on the lectures but a succession of severe strains used me up so completely that there is little prospect of picking up again.

This letter was followed a month later by one from his wife Alice to Mary Herrick in Granville:

Clarence is much better of the heart attack which caused so much anxiety to us all. The night before he gave up the return to Ohio he did not think he should live through the heart disturbance. His trunk was mostly packed and I felt dreadful to give it up after having expected to go and it was so late to ask Charles to take up his work. I think it was best as it came but it was a hard struggle.

There was disappointment in Granville, and some of the people there were critical of his failure to come back after he had had two years of leave on salary. It is not clear how Herrick became aware of the criticism, but he responded somewhat hysterically to it in a letter to his friend Professor George McKibben, on November 17, 1895:

My own life is barren & unfruitful, and much inclined to bitterness. All things by one consent seem to work together for evil and I see neither benevolence nor hope in any perspective. I wish, however, to deny the serious charge of deserting my post from sentimental allurements. On the contrary I would have given a leg to have kept my word and resumed work & now greatly regret that I was frightened out. But the first acute symptoms of heart trouble are alarming and when the heart ceases to beat and the currents cease to move the sequence of sensations is peculiar —to say the least [panic!]. I was ignorant enough to suppose that my days were at an end & so lost my chance. I can now meet the symptoms with indifference which often is tinged with regret that they are not as fateful as I feared at first. I hope you may not know the bitterness of such a disappointment.

Alice Herrick had been less sanguine about her husband's prospects for recovering sufficiently to let him return to his former active life in Ohio, but she did her best to keep him from being depressed. After nine months in the Southwest she had found it

necessary to return and at that time (April 20, 1895) had written
to Dr. Thomas Roberts:

> Dear Friend:
>   Mr. Herrick visited Albuquerque the first of this week for a
> medical examination of his lung and the result was in some ways
> discouraging. We have, therefore, decided to make our home here
> for some years. Perhaps may never be able to return home and it
> is a sad outlook for us all. It is especially trying to Clarence to
> leave his work just as his new laboratories were ready for him to
> do that for which he has been looking for so many years. It is
> necessary for me to leave him for about three months alone after
> which time our family will join us from Ohio. I hope you can
> find time from your busy professional duties to send him an
> occasional letter, for he will be very happy. He values your
> friendship so highly that I feel sure it will greatly cheer him to be
> thus remembered. He feels so sadly cut off from the world.
> P.S.'   Please do not mention my letter.

President Purinton was reluctant to give up hope that Herrick
would return to Denison, but he realized that a salary could not
be provided indefinitely for an absent professor. On June 16,
1896, he told the trustees,

> Professor C. L. Herrick has been in poor health and has been
> absent on sick leave. It was hoped that he might return this year.
> But when the time came, it was not thought safe. During his ab-
> sence, he has finished a course of lectures for his department, and
> has donated periodicals and microscopic preparations valued at
> $400. . . . It is respectfully recommended that for the coming
> year, Prof. C. L. Herrick be relieved of active duty, but retained
> as "Professor in absentia" without pay.

Health was not the only problem plaguing Herrick in New
Mexico. Rarely during the first three years was there enough
money. Some came to him sporadically from Minneapolis and
until June, 1896, there was a generous stipend from Denison
University, but he required still more, for he had acquired heavy
debts in 1892. He tried in various ways to earn money, selling
geological and biological specimens to museums, surveying and
consulting for mining projects, and his wife taught in a private

school for a short time. His knowledge of geology eventually was his salvation. He opened an office for consulting geology and had handbills printed to advertise his service. In October, 1896, he was appointed United States Deputy Mineral Surveyor. Strenuous surveying trips in mountains and desert, and the toll they took on his physical condition, were described at some length by C. J. Herrick (1955). But with all his efforts, he occasionally found himself unable to meet payments on debts and had to call on his brother for help. For example, in early November of 1895 he wrote,

> I find as we are about to go for the mail that we are just about strapped & while I am confidently hoping for money I dare not wait longer. Can you spare us an X for a few days? I do not know how to manage without calling on some of you. Borrow for me from anyone if necessary. I don't know a soul here and don't want to introduce myself by hunting credit. Of course if Minnesota people will send what is due me I shall be all right.

It is hardly thinkable that Herrick under these circumstances could have had inspiration and time to engage in research of any kind, but he found it impossible to remain idle and during the many hours of enforced rest his mind remained active. He retained his editorship and continued to publish editorials, notes, and short articles in the *Journal of Comparative Neurology*. His papers of 1894 were "The Seat of Consciousness" and "Pleasure and Pain." His monograph, with C. H. Turner, "Entomostraca of Minnesota," was transmitted on November 30 to the University of Minnesota by State Geologist Henry F. Nachtrieb, an old friend who had worked under Professors Winchell at Minnesota and Martin at Johns Hopkins. The monograph[1] appeared a year later with a letter of transmittal:

> I have the honor herewith to submit to the honorable Board of Regents my second report . . . together with a report on the Entomostraca of Minnesota by Mr. C. L. Herrick, a graduate of the University and Professor of Biology at Denison University. . . . The report on one group . . . was written by C. H. Turner. . . . These gentlemen have given their services to the survey with-

out charge, having asked for and received barely enough to cover their expenses.

Some of the plates for this publication were drawn by Herrick in New Mexico and engraved by his brother in Ohio.

Denison University retained Herrick on their faculty as professor of Biology for three years—two of them on salary. The only way he could contribute in return was to assemble for that institution a cabinet of geological and biological specimens. The extensive collections, mentioned in President Purinton's report, were housed under the care of Professor Tight in Barney Science Hall. All were destroyed by fire in 1905.

The annual report of President Purinton to the Board of Trustees on July 13, 1897, contained the following information:

> It will be remembered that for some years Professor C. L. Herrick, on account of ill health, has found it impossible to live in Granville. He has been retained as Professor "in absentia," with the hope that he might soon be restored to his work in the University. He has recently become convinced, however, that, for years to come this will not be possible. He wishes me to tender to you his resignation, accompanied with a hearty expression of his high appreciation of the repeated courtesies heretofore shown him by this Board.

A year later he told the trustees: "The ill health of Professor Herrick, his consequent removal to New Mexico and his subsequent election to the presidency of the University of New Mexico may be taken as permanently separating him from the services of Denison University." No one in Granville was more distressed by the loss than President Purinton. Without Herrick there would be no graduate program in biology.

Herrick resigned from Denison University in July, 1897, when he was invited to become the second president of the Territorial University of New Mexico in Albuquerque. He sent news of this to his brother on July 4, 1897:

> Since last writing the unexpected has happened and I am now president of the University of New Mexico with whatever dignity

or disgrace that implies. The salary is only 1800 and I shall have to teach mathematics and pedagogy as well as a little psychology. To Alice this means a great deal as there is a living in prospect for one year at least. The amount of work involved is staggering but I propose to let the other fellow do most of the worry. . . . I hope to live through the ordeal long enough to get out of debt.

The minutes of the Board of Regents of the University of New Mexico meeting of July 3, 1897, noted that the resignation of President E. S. Stover was accepted, whereupon the president of the board nominated Clarence Luther Herrick to succeed him. He was elected, and a salary was authorized to begin on September 1. Herrick did not wait, but started planning at once, and on August 26 a letter from him was read to the Board of Regents. In it he proposed establishment of "a laboratory for microscopical and chemical examination and analysis of morbid products. . . ." No action was taken.

The university consisted of a preparatory department and a normal school housed in a two-story brick building on the outskirts of Albuquerque. Herrick undertook to create academic and graduate divisions, succeeding beyond reasonable expectation in the four years he served as president. He was fortunate in obtaining a faculty; there were several men with academic degrees from eastern universities who had come to Albuquerque for the same reason as Herrick, and he was able to secure their part-time services.

The revised catalog of the University of New Mexico of 1898 listed subjects that were to be given in biological science, noting that the necessary equipment and supplies would be provided from the private laboratory of President Herrick. It did not reveal the deficiency in library facilities—Herrick lacked even a good dictionary, but about this time he obtained his first typewriter. Courses would be given in normal and pathological histology, cellular biology, vertebrate embryology, practical bacteriology, neurology, and technique. The textbook to be used in the biology department was *The Human Body* by H. Newell Martin.

Herrick had two outstanding graduate students in Albuquerque: George E. Coghill in biology and Douglas W. Johnson in geology.

The former came to him soon after he assumed the presidency. The latter arrived in the autumn of 1898. As Coghill held a bachelor of arts degree from Brown University, he was given an appointment as instructor in Solid Geometry in June, 1898, with a salary of seventy-five dollars per month, which was the highest salary paid to any member of the faculty, equaled only by that of the principal of the Normal Department.

When Herrick became president of the University of New Mexico, it seemed appropriate for him to have a more advanced degree than the master's. He submitted a thesis to his alma mater, the University of Minnesota, and was granted the Ph.D. degree in October, 1898. The research on nerve endings in amphibian skin, on which the thesis was based, had been conducted in collaboration with Coghill and published in the *Journal of Comparative Neurology,* as well as in a new periodical in Albuquerque.

Herrick initiated a series of publications known as "Bulletin of the University of New Mexico, Geology Series, and Biology Series." Most of these were articles reprinted from such established journals as the *American Geologist, Bulletin of the Scientific Laboratories of Denison University,* and *Journal of Comparative Neurology.* Those published in 1898 were bound together as Volume I of the "Bulletin of the University of New Mexico" in 1899. A second assemblage was made in 1900, constituting Volume II. The title page of the latter read: "The Bulletin of the Hadley Laboratory, Vol. II, pt. 1, published with the cooperation of Mrs. W. C. Hadley, 1900."[2] There were no more bulletins during Herrick's lifetime.

A solution of the urgent problem of laboratory facilities appeared quite unexpectedly. Mrs. Walter C. Hadley, widow of a prominent citizen of Albuquerque, came to Herrick in January, 1899, with an offer of $10,000 contingent upon $5,000 in contributions from others. The money would build a science hall as a memorial to her late husband. President Herrick announced the gift in an article in the *Mirage,*[3] in which the purpose of the building was disclosed "for scientific study of biological, chemical, physical, and physiological problems relating to the health-giving climate of the New Mexico plateau." Plans of this climatological laboratory building, apparently drawn by Herrick, accompanied the article. Herrick's article on the Walter C. Hadley Science Hall was followed for comparison by one on the Pasteur Institute which had been built in Paris in 1888 at a cost of approximately $500,000.

Herrick entertained hope of developing the university along the lines followed by President Folwell at Minnesota. He wanted to incorporate the Agricultural Experiment Station as well as the School of Mines at Socorro, New Mexico, into a federation of professional schools. Furthermore, he conceived the plan of a marine biological laboratory for research on fresh-water forms to

Herrick's design for the Walter C. Hadley Science Hall at the Territorial University of New Mexico.

round out the academic empire of his dreams. This led to an irrational act that, in view of his past relations with the University of Chicago, is hard to understand.

Herrick wrote to William Rainey Harper in November, 1898, suggesting that the University of Chicago establish a biological research station on Lake Chapola in Mexico. He even proposed that the University of Chicago provide a salaried chair for him, $1,000 for equipment, and a stipend for an assistant! He concluded his appeal with the admonition that he was offering President Harper the "opportunity to make a tardy reparation for the evil I have suffered and to put me right with the scientific world before I die." How could his pride allow him to do such a thing? Harper curtly replied that "the University is not in a position to establish a new chair." He told Herrick the university had given him one professorship which he had resigned "without good reason."

This episode exacerbated Herrick's bitterness and led to outbursts of rage. One of his students of geology recalled many years later that

> in the field when some misunderstanding of agreement on a rendezvous, some undiplomatic remark on my part, or some other minor circumstance provoked him [Herrick] to an explosion of anger which to me was truly terrifying. . . . His face would turn white, and he would pour forth a torrent, never of abuse of me, but of sarcastic abuse of himself as being undoubtedly the guilty party, which however left me in no doubt as to his opinion of my conduct.[4]

Herrick's ill health was assumed to be the cause of these outbursts, and he himself spoke of them as moments of insanity.

His health deteriorated as pressures of the presidency increased, and he spent more time on strenuous expeditions into the mountains. Throughout his long illness Herrick labored under the belief that being out of doors and exercising in the rarified air would be beneficial, whereas rest was what he needed. Finally, on September 17, 1900, he resigned from the presidency, but the board refused to accept his resignation and gave him a year's leave of absence. A successor, his former student and colleague from Denison, Professor William G. Tight, replaced him in the presidency in 1901.

Herrick continued to brood on the Chicago affair and on Harper's curt dismissal of his offer of a way to make amends for the wrong that had been done. On February 9, 1901, he wrote a four-page letter to Professor of Biology C. O. Whitman at the University of Chicago, with the justification that "some facts that have recently become known to me lead me to fear that I have done you (in thought only) an injustice and as the time is now approaching when earthly matters must be brought to 'round-up,' I shall be glad to be set right if the misconception really exists." He proceeded to review the series of events from his first negotiations with Harper to the final resignation in 1892. He told Whitman that although he had not originally blamed Donaldson for accepting the chair of Neurology, "[my] proposal of a little cooperation in Neurology brought me the most insulting letter I ever received from anyone." Apparently Herrick forgot that in April, 1898, he had invited Donaldson to join the editorial board of the *Journal of Comparative Neurology.* His letter to Whitman continued:

> More recently I have learned that Dr. Harper has denied the whole matter, in fact I have a letter in which he so does in spite of the documentary and other proof. It occurs to me that neither you or Donaldson really knew that I had received the most solemn assurances of the free hand in these departments and that you may have been deceived by Dr. Harper. I dislike to carry hence a false impression regarding scientific men for whose work and character I have only admiration except in the very specific matter I have mentioned. The bitterness of the injury is responsible for the disease that prevented me from carrying out my life work and has made me an outcast for seven years. . . . I dislike to die with the knowledge that my good name has been injured in the house of those who should have been my friends by such a scoundrel as Harper. For the religion illustrated by such folk I can [say] less than nothing but still believe in honor.

Herrick begged Whitman to reply, but met only silence.

He waited a year and then prepared a document, labeled

"Statement of C. L. Herrick" which, after a preamble, continued as follows:

> This statement relating to the circumstances which led to my withdrawal from the University of Chicago after having been elected full professor of Biology therein, is intended . . . solely to protect my children from the further effect of the conduct mentioned. I therefore present the facts as they occurred and my sworn testimony thereto.

The document goes on for four pages and concludes with the following paragraph:

> The only subsequent communication from Dr. Harper denied the above facts in toto. Events have proven the wisdom of my withdrawal, though it is not without bitterness that I realize that supposed friends now in that institution have been estranged by false representations from the president of that institution. Accordingly, the foregoing brief statement is made in the belief that it may at least relieve my memory from the insult that I must in the present suffer.

The statement was not notarized and remains unsigned among the private papers of C. L. Herrick at the University of Kansas.

The bibliography of Clarence Luther Herrick for the ten years in New Mexico contains only two titles of articles based on original research in the nervous system. He published sixty-six articles of one kind or another, including geological and philosophical ones, but lacked the facilities to continue with neuroscience even had his health permitted him to do so. That did not mean that he ceased to think about the brain and its functions.

The files of his papers at the University of Kansas contain some unpublished manuscripts, the most important one of which is entitled "Textbook of Comparative Neurology." The manuscript is in rough form, much of it in longhand on stationery of the University of New Mexico, which indicates that he wrote most of it after he became president of that institution. It is filled with preliminary sketches for illustrations he proposed to make later. It contains other drawings from previous publications and some newly finished unlabeled ones. There are photographs of a human brain, printed on blueprint paper.

The proposed textbook covered many topics in neurology—cerebral localization, trophic function of nerves, the hypophysis, etc. It was written from embryological as well as comparative points of view. Development was illustrated by drawings of a 26-mm human fetus and sections through it. The book could appropriately have been called "Comparative and Developmental Neurology." Had Herrick lived to finish this work, it would have assured him a place at the forefront of American neurologists, for there was no textbook of the kind at that time. It could have been (but probably was not) an inspiration for his brother's book, published eleven years after Clarence's death.[5]

Herrick published a lengthy account entitled "Development of the Brain" in the second edition of Buck's *Reference Handbook of the Medical Sciences* in 1901. It was based largely upon writings of European neuroembryologists (Balfour, His, Kolliker, Mihalkovics, Selenka). The subject matter was morphological and comparative, containing little in the way of intrinsic structure. The article was illustrated with forty-eight drawings, most of which were borrowed from other authors. Apparently Herrick intended to incorporate this in his "Textbook of Comparative Neurology." It bears testimony to his abiding interest in embryology that dated back to his days in the Young Naturalists' Society.

Herrick did not neglect his interest in psychology. He left the major part of a textbook of psychology, typed on stationery of the University of New Mexico. Another manuscript, apparently completed and nearly ready for a publisher, was called "Lectures on Conduct. The Principles of Ethics from the Dynamic Point of View." In the preface he noted that "the presentation is avowedly from the standpoint of 'Functional Psychology' and 'Dynamic Monism' in philosophy. The doctrine of evolution is assumed throughout but no one of its modern 'ism.' The point of view is scientific rather than dogmatic." A former student and colleague at Denison, H. H. Bawden, compiled and edited some of Herrick's unpublished fragments on philosophical and psychological topics in 1910.[6]

Herrick maintained an interest in the studies being carried on in Granville by his brother Charles, and on August 5, 1904, he wrote to him to say that he was

> glad to know that something is coming out of animal behavior. You know my plans were to fill our yard with kennels and hutches and to keep that side going along with the anatomical. The "Psychic Life of Microorganisms" set a good pace for some phases of the work. In my western experience I am impressed with the utter stupidity and nongeneralizable nature of reaction to unexpected conditions.

Herrick bought an adobe house in the old Spanish town of Socorro, New Mexico Territory, intending to make it his permanent residence. The deed, recorded on 25 April 1904, indicates that he paid "one hundred and ten dollars lawful money to F. P. Shaw and wife" for the property. Apparently the Herrick family were occupying the house before the deed was recorded because he mentioned it earlier in a letter to his mother and drew for her a picture of the fireplace he had added.

The last letter written by Herrick to his mother on March 5, 1904,[7] contains comments on his manuscript on ethics; in it he summarized the final state of his moral philosophy and his opinion of the body-soul concept:

> I have wondered a little whether you were able to get the drift of my ethics MS. . . . The underlying principle is the unity of the cosmos and the belief that realities are active. That only what is doing something really exists. In seeking for motives for our acts evolutionists show that the earliest and most persistent impulses are those which tend to preserve and perfect self. It is not presumed that the Creator made any mistake in this. My work was to indicate that moral living did not repudiate these laws but extended them. . . . Another thing which I hoped to make more plain is a truth the failure to recognize which has been responsible for much of my suffering and failure. I used to believe that man "is a soul and has a body" and I thought the tool was a poor one at best and I could use it as long as I could and

throw it away afterward. I am now convinced that the body is just as much an expression of the soul as thought is and that the being is all one. One cannot neglect any part of the self without suffering. If I had learned to care for the body as well as I now could 20 years ago I might have 20 years more.

To the very end, Herrick retained his interest in the nervous system. That he was able to do so little about it is explained in a letter to his brother in midsummer of 1904, in which he said,

Am recovering from our latest unpleasantness and taking life easy. Am likely to have to sink a well for coal exploration, superintend building a small railroad, and survey a mine for patent this fall. Wish all this work were in Tophat [Topheth?], but we must eat. I could do a little neurology if I dared let things slide.

Clarence Herrick was invited to speak at the International Congress of Arts and Science that was to be held in conjunction with the Louisiana Purchase Exposition in St. Louis, Missouri. He wrote to his brother in August, 1904:

I received notification that in event of my being unable to be in St. Louis it would be agreeable to have my message delivered by you. . . . I am sending a short communication and if it is convenient for you to see Professor Sanford and read it at the proper session of the section of Comparative and Genetic Psychology I shall be pleased. The idea is to get a bit of free advertising for the *Journal* in as many places as possible and my name was used in that connection.

Charles Judson Herrick read the paper on September 24, 1904, concluding with his brother's benediction:

Let us hope that the comparative psychologist of the future will not disdain to follow the latest work in histology. We may be proud that one of our own genetic psychologists has made substantial contribution to evolutionary theory; and when biologists condescend to cooperate with comparative psychologists, both sciences may begin to "think in the round."

At the age of forty-six years, Clarence Luther Herrick's death had occurred on September 15, 1904, in Socorro, New Mexico. He was buried there in the Socorro Cemetery.

NOTES

1. C. L. Herrick and C. H. Turner, Synopsis of the Entomostraca of Minnesota, with a description of related species, comprising all known forms from the United States included in the orders *Copepoda, Cladocera, Ostracoda. Geol. Nat. Hist. Survey Minn., Zool. Ser.* 2: 1-525, 81 pls., 1895.
2. Copies of these two volumes are in the Denison University Archives.
3. C. L. Herrick, An opportunity and a responsibility. *The Mirage* 1 (3): 3-5, 1899.
4. Letter of D. W. Johnson to C. J. Herrick, October 1, 1939 (Kansas file).
5. C. J. Herrick, *An Introduction to Neurology* (Philadelphia: Saunders, 1915).
6. H. H. Bawden, The metaphysics of a naturalist. Philosophical and psychological fragments by the late C. L. Herrick. *Bull. Sci. Lab. Denison Univ.* 15: 1-99, 1910.
7. Anna Strickler Herrick recovered from her psychosis after twenty-five years and lived to be almost 105 years old.

XII

PERSONALITY

Several aspects of the life of Clarence Luther Herrick have been described by people who knew him, but only his brother Charles Judson attempted to write a complete biography. Others differed in their point of view, memorializing him as a naturalist, geologist, or teacher and recalling only limited periods of his life. Lincoln Blake, for example, did not know him, but explored the Harper-Herrick relationship in some depth as it concerned the development of science in the University of Chicago. George Coghill's and Douglas Johnson's commentaries were limited to their student days in New Mexico. Alfred Cole and William Tight knew him for longer periods at Denison University. None of these men could have accounted for Herrick's entire life.

Obituaries and biographical sketches prepared for commemorative occasions tend to emphasize qualities of character that the writers hope will glorify the memory of a person, and not infrequently some unflattering details are left out. This was true in the biography written by Herrick's brother. The files of letters in the Kenneth Spencer Library at Lawrence, Kansas, reveal pertinent matters that were omitted.

It is understandable that Charles Judson Herrick's reverential attitude toward his brother guided his pen. Clarence was his childhood teacher, supervising home chores and keeping him out of mischief. A different relationship developed after their father's death in 1886, when Charles Judson's juvenile hero worship of his elder brother began to be superseded by a professional admiration coupled with deep affection in later years, making it difficult for him to prepare an unbiased account of his brother's life.

He wrote and, on more than one occasion, spoke about his brother's "life plan." But nowhere did I encounter a plan spelled

121

out by Clarence Herrick himself. He was, like most of us, an opportunist. His sudden inspiration to switch from studying the circulation of *Daphnia* to the psychology of Lotze is a case in point. Until he met Miss Elizabeth Hamilton in Leipzig he had formulated no plan to explore the brain as a way toward understanding the soul. Soon thereafter he found it expedient to engage in geology in central Ohio and he put aside his thoughts of psychology and his work on the mammals of Minnesota to collect rocks. Four years later he abandoned geology and plunged into neurobiology. As the opportunities appeared, Herrick grasped them.

Charles Judson Herrick did not portray his brother's impetuosity—one of Clarence's most pronounced characteristics—as fully as he might have done. Time and again, Clarence made decisions based on impulsive judgments. For example, he suddenly abandoned his new house in Granville when emotional instability made it uncomfortable to remain there. He audaciously launched publication of the *Journal of Comparative Neurology* without giving the problem careful consideration. Another example—a brilliant one, to be sure—was his acceptance of George E. Coghill as a graduate student, although the latter had no qualifications for the work he proposed to do and Herrick could offer practically no facilities with which to do it.

Young Coghill had graduated from Brown University with a bachelor of arts degree, after which he had tried to pursue graduate work in a theological seminary. He soon found himself in such a conflict with principles of his fundamentalist teachers that he had to abandon his studies. Emotionally disturbed, he retired to the mountains of New Mexico to meditate and contemplate his future and in Albuquerque he met Herrick who had become president of the University of New Mexico. He talked to him and swiftly developed an interest in "the nervous system as an approach to psychology and philosophy," whereupon he was given access to Herrick's laboratory and to his library, such as it was.

Later Coghill declared that "Professor Herrick at once became my inspiration; and as such he still lives in all I do. He lifted me from the 'Slough of Despair' and placed me on the highest ground of hope and inspiration." Perhaps Herrick recognized in Coghill a troubled mind not unlike his own.

After the death of Clarence Luther Herrick, the loyalty he had felt toward this student was assumed by Charles Judson Herrick. A close friendship led to Coghill being chosen in 1927 as the managing editor of the *Journal of Comparative Neurology*. The memoirs of both Charles Judson Herrick and George Coghill express their deep personal feelings for their teacher.

William Tight was the closest to Herrick of all his students at Denison University and later succeeded him as president of the University of New Mexico. He had been one of the first two students to earn the master of science degree under Herrick. He contributed an obituary to the *Bulletin of the Scientific Laboratories* in 1905 in which he said:

> I know of no better word to express the general characteristic of the man than one which I have heard often used in reference to him, and which he has used often to me in reference to himself, and that is "pioneer." A pioneer in every sphere of his activity, it was his task to lay foundations among the difficulties.

Several of Herrick's personal qualities, especially his leadership and inspirational teaching, have been stressed by other writers. He was a leader rather than a follower. He showed a strong tendency to stand alone, but did not avoid group activities. He seldom sought advice from his colleagues, an exception being that of his physician friend, Thomas Roberts, to whom he turned when illness struck. He was embarrassingly outspoken at times. With a measure of conceit, he recognized his own genius.

Rarely was Clarence Luther Herrick completely idle, for inactivity was something he could not endure. This was true in his youth when he guided the affairs of the Young Naturalists' Society, and it was true even during his time of terror when he believed himself to be near death. He lived his life with an intensity that few other scientists have equaled. Tasks, in no matter what field of endeavor, were undertaken with furious drive. Note his productivity of profusely illustrated research publications during the Cincinnati period: Twenty articles based on his neurological research appeared between 1889 and 1892.

Herrick made no great and lasting scientific discovery and many of his publications were fragmentary, but he left as a monument the *Journal of Comparative Neurology*. Notably, he developed the concept of structural and functional integration in the nervous system, a concept that he imparted to his worshipful student, George Coghill. Moreover, he stimulated the emergence of physiological psychology as a science. These were among his best contributions.

Herrick's first thirty-four years were filled with the pleasure and excitement of accomplishment. There were frustrations, as in Granville in 1888, but he circumvented most of them and went forward with his work. Generally he was at peace with God and his fellowmen. He viewed the future with implicit confidence as he laid plans for a neuroscience (he did not use the term) that would combine the technique of histology, embryology, and physiology of the nervous system with those of comparative psychology. Had he been able to implement his plan, the result would have stood as his greatest achievement.

Herrick influenced the scientific careers of numerous students at Denison, Cincinnati, and New Mexico, but eminence in neurology was attained only by George Coghill and Charles Judson Herrick. The latter succeeded him in comparative neurology, but he did not follow through with his older brother's plans for an integrated neuroscience program. Another brother, William, was uninfluenced by Clarence's work in neurology. His son, Harry, did follow his father's geological interests.

Charles Judson Herrick assumed management of the *Journal of Comparative Neurology* in 1894 when Clarence became ill. That and teaching duties along with his own graduate studies imposed a heavy burden. He did extraordinarily well to forge his own more limited career in comparative neurology.

Looking through the files of unpublished manuscripts of Clarence Luther Herrick, one wonders why Charles did not try to complete some of them. For example, why did he not undertake to finish the "Textbook of Comparative Neurology"? Eleven years after his brother's death, Charles Judson Herrick did publish his own *Introduction to Neurology,* which bears no relationship to Clarence's manuscript; only two of Clarence's publications (one posthumous) were cited by his brother in that book.

The year 1892 marked an abrupt change in Clarence Luther Herrick's personality. From the pinnacle of success, he saw his hopes and aspirations dashed; frustration beyond comprehension struck him, shattering his faith in human beings and their religious principles. To cap all this, he lost his health and was forced into periods of desperate irrationality. The last ten years of his life were tragic years in spite of accomplishments in New Mexico. One cannot read his letters without sensing the utter devastation of his defeat. His pleas to colleagues for sympathetic understanding were unanswered. Few, save his own family including his brother Charles Judson and his student George Coghill, deeply mourned his passing, for in ten years he had passed out of activity in science and his contributions were but dimly remembered.

APPENDIX I

HERRICK'S PUBLICATIONS IN NEUROSCIENCE

This list contains Herrick's articles in neuroscience published in scientific journals and books. It includes some essays of a philosophical nature but excludes his writings that appeared in religious magazines, few of which were relevant to neuroscience. The titles were taken from the biography published by Charles Judson Herrick (1955), which included nonneurological articles.

1. *Outlines of Psychology: Dictations from Lectures by Hermann Lotze.* Translated with the addition of a chapter on the anatomy of the brain. Minneapolis: S. M. Williams. x+150 pp., 2 pls., 1885.
2. A contribution to the histology of the cerebrum. *Cincinnati Lancet-Clinic, N. S.* 23: 325-27, 1889.
3. Notes upon the brain of the alligator. *J. Cincinnati Soc. Nat. Hist.* 12: 129-62, 9 pls., 1890.
4. Suggestions upon the significance of the cells of the cerebral cortex. *The Microscope* 10: 33-38, 2 pls., 1890.
5. The central nervous system of rodents. Preliminary paper. (With W. G. Tight.) *Bull. Sci. Lab. Denison Univ.* 5: 39-95, 19 pls., 1890.
6. The evolution of the cerebellum. *Science* 18: 188-89, 1891.
7. The commissures and histology of the teleost brain. *Anat. Anz.* 6: 676-81, 1891.
8. Contributions to the comparative morphology of the central nervous system. I. Illustrations of the archetectonic of the cerebellum. *J. Comp. Neurol.* 1:5-14, 4 pls., 1891.
9. Contributions to the comparative morphology of the central

126

nervous system. II. Topography and histology of the brain of certain reptiles. *J. Comp. Neurol.* 1: 14-37, 2 pls., 1891.
10. The problems of comparative neurology. *J. Comp. Neurol.* 1: 93-105, 1891.
11. Contributions to the comparative morphology of the central nervous system. III. Topography and histology of the brain of certain ganoid fishes. *J. Comp. Neurol.* 1: 149-82, 4 pls., 1891.
12. Neurology and psychology. *J. Comp. Neurol.* 1:183-200, 1891.
13. Metamerism of the vertebrate head. *J. Comp. Neurol.* 1: 203-4, 1891.
14. Contributions to the morphology of the brain of bony fishes. Parts I & II. (With C. J. Herrick.) *J. Comp. Neurol.* 1: 228-45, 333-58, 5 pls., 1891.
15. Notes upon the anatomy and histology of the proscencephalon of teleosts. *Amer. Nat.* 26: 112-20, 2 pls., 1892.
16. Additional notes on the telost brain. *Anat. Anz.* 7: 422-31, 10 figs., 1892.
17. Notes upon the histology of the central nervous system of vertebrates. *Festschrift zum siebenzigsten Geburtstage Rudolf Leukarts,* 278-88, 2 pls., Leipzig: W. Englemann, 1892.
18. The cerebrum and olfactories of the opossum, *Didelphys virginica. J. Comp. Neurol.* 2: 1-20, 1892; and *Bull. Sci. Lab. Denison Univ.* 6: 75-94, 3 pls., 1892.
19. Contribution to the morphology of the brain of bony fishes. Part II. Studies on the brain of some American fresh-water fishes (continued). *J. Comp. Neurol.* 2: 21-72, 8 pls., 1892.
20. Neurologists and neurological laboratories. No. 1. Professor Gustav Fritsch. With portrait. *J. Comp. Neurol.* 2: 84-88, 1892.
21. The psychophysical basis of feelings. *J. Comp. Neurol.* 2: 111-14, 1892.
22. Instances of erroneous inference in animals. *J. Comp. Neurol.* 2: 114, 1892.
23. Instinctive traits in animals. *J. Comp. Neurol.* 2: 115-36, 1892.

24. Histogenesis and physiology of the nervous elements. *J. Comp. Neurol.* 2: 137-49, 1892.
25. Intelligence in animals. *J. Comp. Neurol.* 2: 157-58, 1892.
26. Embryological notes on the brain of the snake. *J. Comp. Neurol.* 2: 160-76. 5 pls., 1892.
27. Localization in the cat. *J. Comp. Neurol.* 2: 190-92, 1892.
28. Comparative psychology. *J. Comp. Neurol.* 2: i-iii, 1892.
29. The scope and methods of comparative psychology. *Denison Quarterly.* 1: 1-10, 134-41, 179-87, 264-81, 1893.
30. Nervous system: Comparative anatomy, 669-91. In: *Reference Handbook of the Medical Sciences, Supplement,* Vol. 9. A. H. Buck [Ed.]. New York: William Wood, 1893.
31. Nervous system: Histogenesis of its elements, 691-96. In: *Reference Handbook of the Medical Sciences, Supplement,* Vol. 9. A. H. Buck [Ed.]. New York: William Wood, 1893.
32. Emotions, the physiological and psychological basis of the, 269-73. In: *Reference Handbook of the Medical Sciences, Supplement,* Vol. 9. A. H. Buck [Ed.]. New York: William Wood, 1893.
33. Waller's law, 996-97. In: *Reference Handbook of the Medical Sciences, Supplement,* Vol. 9. A. H. Buck [Ed.]. New York: William Wood, 1893.
34. The evolution of consciousness and of the cortex. *Science,* 21: 351-52, 1893.
35. The development of medullated nerve fibers. *J. Comp. Neurol.* 3: 11-16, 1893.
36. The scientific utility of dreams. *J. Comp. Neurol.* 3: 17-34, 1893.
37. The hippocampus in Reptilia. *J. Comp. Neurol.* 3: 56-60, 1893.
38. Contributions to the comparative morphology of the central nervous system. II. Topography and histology of the brain of certain reptiles (continued). *J. Comp. Neurol.* 3: 77-106, 119-40, 11 pls., 1893.
39. Report upon the pathology of a case of general paralysis. *J. Comp. Neurol.* 3: 141-62, 1893; and *Columbus State Hospital for the Insane,* Bulletin No. 1, 5 pls., 1893.

40. The callosal and hippocampal region in marsupial and lower brains. *J. Comp. Neurol.* 3: 176-82, 2 pls., 1893.
41. The soul of man. *J. Comp. Neurol.* 3: lxvi-lxviii, 1893.
42. The seat of consciousness. *J. Comp. Neurol.* 4: 221-26, 1894.
43. Pleasure and pain. *J. Comp. Neurol.* 4: lxxxvi-lxxxii, 1894.
44. The histogenesis of the cerebellum. *J. Comp. Neurol.* 5: 66-70, 1894.
45. Notes on child experiences. *J. Comp. Neurol.* 5: 119-23, 1895.
46. The cortical optical center in birds. *J. Comp. Neurol.* 5: 208-9, 1895.
47. Neurology and monism. *J. Comp. Neurol.* 5: 209-14, 1895.
48. Mental development of the child. *J. Comp. Neurol.* 5: xxiv-xxxvii, 1895.
49. The present state of comparative psychology. *J. Comp. Neurol.* 5: xliii-xliv, 1895.
50. Suspension of the spatial consciousness. *Psychol. Rev.* 3: 191-92, 1896.
51. Focal and marginal consciousness. *Psychol. Rev.* 3: 193-94, 1896.
52. The testimony of heart disease to the sensory facies of the emotions. *Psychol. Rev.* 3: 320-22, 1896.
53. Illustrations of central atrophy after eye injuries. *J. Comp. Neurol.* 6: 1-4, 1896.
54. Lecture notes on attention. An illustration of the employment of neurological analogies for psychical problems. *J. Comp. Neurol.* 6: 5-14, 1896.
55. The psycho-sensory climacteric. *Psychol. Rev.* 3: 658-61, 1896.
56. Is the decorticated dog conscious? *J. Comp. Neurol.* 6: xxi-xxiii, 1896.
57. Psychological corollaries of modern neurological discoveries. *J. Comp. Neurol.* 7: 155-61, 1897.
58. Inquiries regarding current tendencies in neurological nomenclature. (With C. J. Herrick.) *J. Comp. Neurol.* 7: 162-68, 1897.
59. The propagation of memories. *Psychol. Rev.* 4: 294-96, 1897.

60. Physiological corollaries of the equilibrium theory of nervous action and control. *J. Comp. Neurol.* 8: 21-31, 1898; and *Bull. Univ. of New Mexico, Biol. Ser.* 1 (1), 1898.

61. The somatic equilibrium and the nerve endings in the skin. Part I. (With G. E. Coghill.) *J. Comp. Neurol.* 8: 32-56, 1898; and *Bull. Univ. of New Mexico, Biol. Ser.* 1 (2), 1898.

62. The cortical motor centers in lower mamals. *J. Comp. Neurol.* 8: 92-98, 1898.

63. The vital equilibrium and the nervous system. *Science* 7: 813-18, 1898.

64. Substitutional nervous connection. *Science* 8: 108, 1898.

65. The material versus the dynamic psychology. *Psychol. Rev.* 6: 180-87, 1899.

66. Clearness and uniformity in neurological descriptions. *J. Comp. Neurol.* 9: 150-52, 1899.

67. Brain weight and mental capacity. *J. Comp. Neurol.* 10: iii-v, 1900.

68. Nervous system. (With C. J. Herrick.), 153-66. In: *Dictionary of Philosophy and Psychology,* Vol. 2. J. M. Baldwin [Ed.]. New York: Macmillan, 1901 (new edition, 1925).

69. Brain: development of the, 268-82. In: *Reference Handbook of the Medical Sciences,* 2d ed. Vol. 2. A. H. Buck [Ed.]. New York: William Wood, 1901.

70. Nervous end organs, 818-25. In: *Reference Handbook of the Medical Sciences,* 2d ed. Vol. 2. A. H. Buck [Ed.]. New York: William Wood, 1901.

71. Color vision (a critical digest). *J. Comp. Neurol.* 14:274-80, 1904.

72. Recent contributions to the body-mind controversy. *J. Comp. Neurol.* 14: 421-31, 1904.

73. The law of congruousness and its logical application to dynamic realism. *J. Philos.* 1: 595-603, 1904.

74. Mind and body—the dynamic view. *Psychol. Rev.* 11: 395-409, 1904.

APPENDIX II

HERRICK'S PROPOSAL FOR STUDIES IN NEUROSCIENCE AT CHICAGO

GRADUATE DEPARTMENT OF NEUROLOGY AND COMPARATIVE PSYCHOLOGY

In accordance with the general plan for graduate instruction and investigation, it is hoped to present in this department the necessary facilities for investigating the anatomy and physiology of the nervous system to the fullest extent possible. Approach is opened from all accessible aspects of this many-sided subject.

It is needless to state that such study is essentially and primarily comparative, and no fact which bears however indirectly upon nervous function is excluded. The term comparative psychology is used, regardless of those questions which might be raised as to the validity of psychological data from lower animals, to cover all phenomena which resemble or are analogous with the expression of emotion or intelligence as exhibited by all animals. Physiological psychology is distinctly included and is considered but another phase of a truly comparative science of neurology. The rational psychology is, however, only discussed in so far as it is a necessary preliminary to what is pure phenomenal.

The special organ of this department will be the *Journal of Comparative Neurology,* though as hitherto it will be open to contributions from all sources and will contain as full a résumé of literature and technique as possible.

In recognition of the fact that most of the important additions to our knowledge of the brain have been derived from anatomical investigation or experiment dependent thereon, all work in this department will suppose a thorough preliminary investigation of the structure of the nervous system and every facility will be extended to those who may desire to apply modern technique to the solution of the vexed questions of neurohistology and morphology. Those electing work in any direction within this department will ordinarily be expected to spend some time in the [course in Comparative Neurology].

Course in Comparative Neurology

I. Didactic
1. Lectures on the comparative morphology of the brain of vertebrates.
2. Course on the structure of the human brain based on Edinger.
3. Lectures on experimental neurology with illustrative experiments and study of autopsies.

II. Laboratory
1. Laboratory practicum in comparative morphology of the vertebrate brain. The student will prepare and study minutely brains and sections of brains of fishes, amphibians, reptiles, birds and a mammal.
2. Students will undertake experimental work in lines suggested by the department.
3. Seminary. The neurological seminary, embracing all departments of the work, will meet weekly for informal comparison of notes.

III. Independent Research.
   After one year or less of such work as above indicated the student may elect to pursue some line of morphological investigation involving a study of some problem in the anatomy of the nervous system.

Course in Physiological Psychology

Students adequately prepared in physiology as taught in the

college course and who shall have pursued the lecture courses indicated above under 1 may enter a course in Physiological Psychology as follows:

I. Didactic
   1. Lectures on physiological psychology with reference to Wundt, Ladd, and Ziehen.
   2. Special lectures in optics and acoustics with physical illustrations and experiments.
   3. A brief course of lectures on the history of opinion relating to the relation of body and soul, etc.

II. Practical
   1. A laboratory course in the morphology and histology of the organs of sense and their development in certain groups.
   2. An experimental course in psycho-physics—with a repetition of the classical experiments respecting the nature and rate of propagation of nerve stimuli and the phenomenon of sensation.
   3. Special studies in the relation of physical to psychical manifestations.

III. Independent Research
   Those sufficiently prepared in rational psychology as well as the above topics will be encouraged to undertake the solution of special questions of physiological psychology.

Course in Comparative Psychology

This course presupposes considerable work in both preceding departments and is believed to be unique in scope and nature.

I. Didactic
   1. Lectures in the expressions of emotion in animals with a discussion of the validity and limitations of such data.
   2. Careful compilation and criticism of the materials of the science as presented by Darwin, Romanes, Morgan, and others.

II. Practical
   The work in this direction will be adapted to existing

conditions. It will be attempted to apply a truly experimental method to the study of the psychical expressions of animals; e.g., circumstances being varied so as to presumably produce a given emotion or psychical activity the animal will be photographed by means of a camera with a slowly rotating multiple diaphragm thus producing a series of photographs of the animal separated by a few seconds or, in other words, a complete photographic record of the changes in aspect or movements evoked. A comparison of the reaction of different animals when affected by the same emotion or affection may be attempted.

In connection with this department it is hoped to establish a statistical bureau for the collection and critical study of the physical, nervous, and mental development of schoolchildren. Data of this sort will be most valuable in the settlement of numerous questions relating to heredity as well as of vast importance to the science of pedagogy.

BIOGRAPHICAL REFERENCES

Sources of material on the life of Clarence Luther Herrick were mentioned in the Preface. Charles Judson Herrick published the only full biography. It and other principal biographical articles are listed below and are referred to in chapter notes by name and date.

Bawden, H. H.   Clarence Luther Herrick. *J. Comp. Neurol.* 14: 515-34, November, 1904.

Blake, L. C.   The Concept and Development of Science at the University of Chicago 1890-1905. Doctoral Dissertation, University of Chicago, September, 1966.

Cole, A. D.   Clarence Luther Herrick. *Science* 20: 600-601, November 4, 1904.

Cole, A. D.   C. L. Herrick as a maker of scientific men. *Bull. Sci. Lab. Denison Univ.* 13: 1-13, 1905.

Herrick, C. J.   Clarence Luther Herrick: Pioneer naturalist, teacher, and psychologist. *Trans. Amer. Phil. Soc. N. S.* 45 (1): 1-85, March, 1955.

Tight, W. G.   Clarence Luther Herrick. *Amer. Geol.* 36: 1-26, July, 1905.

Tight, W. G., C. E. Hodgin, and J. Weinzirl.   In memoriam. *Univ. New Mex. Weekly* 7 (4): September 24, 1904.

Windle, W. F.   Clarence Luther Herrick and the beginning of neuroscience in America. *Exp. Neurol.* 49 (1, pt. 2): 1-10, 1 pl., 1975.

INDEX

Agnosticism, 74
Alabama, 42, 53
Albuquerque, 106, 111
Alligator, 60
Alumni, Chicago, 99
Ambidexterity, 22
American Association for Advancement of Science, 17, 61
American Journal of Neurology, 16
American Men of Science, 43
American Naturalist, 16
American Neurological Association, 17
American Physiological Society, 17
Anatomische Anzeiger, 78
Anderson, Galusha, 54, 56
Animalcules, 27
Animals, experimental, 60
Anxiety, 56
Architect, 54
Army, 22
Arno Press, 42
Association of Colleges of Ohio, 49
Association of American Anatomists, 17, 62

Baptist, 20, 45, 74
Barney, Eugene J., 55, 97
Barney Science Hall, 104
Bawden, H. H., 49, 117
Behrens, Adolphus, 102
Berlin, 65, 76, 78, 90*ff*
Bible reading, 58
Biology: at Chicago, 84; at Cincinnati, 59; at Denison, 97

Biology vs. religion, 74
Blake, Lincoln, 121
Body-soul, 118
Bowditch, H. P., 17
Brain: alligator, 60; avian, 62; early interest, 26; groundhog, 60; reptile, 94; structure, 60
Brothers, 23
Brown-Séquard, C. E., 15
Building, science, 53, 55, 113
Bulletin of the Scientific Laboratories of Denison University, 50, 65, 97
Bulletin of the University New Mexico, 112

Camping, 25
Cell, pyramidal, 60
Cemetery, 120
Cerebral cortex, 54
Cerebral localization, 59
Chapel, 28
Chicago, 73
Cheney, J. L., 87
Cincinnati College, 58
Clarence, N.Y., 19
Classics of Psychology, 42
Coghill, George E., 112, 121*ff*
Cole, Alfred, 47, 98, 121
Congress of Physiological Sciences, 17
Cortical localization, 62
Cortical stimulation, 61
Crustacea, 33, 53
Curriculum: at Denison, 99; at New Mexico, 111

Language, German, 23
Latin School, 28
Leave, 35, 56, 103, 106
Leipzig, 16, 35
Leukarts, Rudolf, 35, 93
Libraries: Athenaeum, 30; Cincinnati, 58; Denison, 52, 57, 96; Granville, 52; Kenneth Spencer, 121; Minnesota, 28, 31
Licking County, 49
Lithography, 34, 65
Lithology, 52
Loeb, J., 81
Lotze, Rudolf Hermann, 36*ff*

McKibben, George, 98, 107
McMicken Review, 59
Malaria, 53
Mammologist, 33
Marriage, 42
Martin, H. Newell, 17, 44, 111
Massacre, 30
Medical courses, 58, 99
Melodeon, 21
Memoirs of the Denision Scientific Association, 53
Meyer, Adolf, 93
Microphotography, 34
Microscope, 27, 48, 65
Microscope, 16
Microtome, 52
Mikrokosmus, 44
Minneapolis, 19*ff,* 104*ff*
Minnesota Plan, 28
Mirage, 112
Monism, 103
Museums: Denison, 106; Minneapolis, 25
Music, 21

Nachtrieb, H. F., 109
Naturalist, 25, 33
Nerve cell, 60
Neurobiology, 59

Neuroembryology, 71
Neuroglia, 61
Neurohistology, 62
Neuropathology, 61
Neuroscience, 15, 76, 103, 124, 131
Newark, 54
New Mexico, 105*ff*
New York City, 19

Ontology, 102
Ornithology, 26
Outlines of Psychology, 39, 41
Owen, Alfred, 56

Parents, 20
Periodicals, exchange, 52, 61, 65, 96
Philadelphia, 61
Philosophy, 85, 89
Physiology, 54, 61
Popular Science Monthly, 16
Presidency, 110
Princeton, 62, 94
Printer, 67, 68, 93
Professor: in absentia, 108; biology, 93, 97; geology and natural history, 45; neurology, 78, 84
Professorships at Chicago, 64, 78
Psychology, 16, 54, 76, 84, 102
Psychomotor, 60
Psychosensory, 60
Purinton, Daniel B., 63, 96*ff,* 106, 110

Ramón y Cajal, S., 15
Regeneration, 63
Religion, 20; vs. biology, 74
Residence, 55, 64, 118
Resignation: Chicago, 81; Cincinnati, 64; Denison, 56, 110; New Mexico, 114; State Mammalogist, 43
Rhodes, W. C. P., 47
Rivers: Mississippi, 19; Ohio, 63; St. Louis, 42